BUSINESS ISSUES, COMPETITION AND ENTREPRENEURSHIP

SMALL BUSINESS INNOVATORS

INSIGHTS FROM ACCELERATORS, ADDITIVE MANUFACTURING AND SUPPLY CHAIN ANALYSIS

BUSINESS ISSUES, COMPETITION AND ENTREPRENEURSHIP

Additional books in this series can be found on Nova's website under the Series tab.

Additional e-books in this series can be found on Nova's website under the e-book tab.

BUSINESS ISSUES, COMPETITION AND ENTREPRENEURSHIP

SMALL BUSINESS INNOVATORS

INSIGHTS FROM ACCELERATORS, ADDITIVE MANUFACTURING AND SUPPLY CHAIN ANALYSIS

MARY BROWN
EDITOR

nova publishers
New York

NOTICE TO THE READER

Additional color graphics may be available in the e-book version of this book.

Library of Congress Cataloging-in-Publication Data

ISBN: 978-1-63483-265-6

Published by Nova Science Publishers, Inc. † New York

CONTENTS

Preface		**vii**
Chapter 1	Small Innovative Company Growth—Barriers, Best Practices and Big Ideas: Lessons from the 3D Printing Industry *Mark Harrison*	**1**
Chapter 2	Supply Chain Innovation: Strengthening America's Small Manufacturers *Executive Office of the President and the U.S. Department of Commerce*	**69**
Chapter 3	Innovation Accelerators: Defining Characteristics among Startup Assistance Organizations *C. Scott Dempwolf, Jennifer Auer and Michelle D'Ippolito*	**93**
Index		**143**

PREFACE

Innovation in the United States has been one of the driving forces in our development as one of the leading economies in the world. Small companies comprise the overwhelming majority of all businesses in the United States and they must be able to effectively and efficiently bring their innovative products and services to market and grow. In this book, the U.S. Small Business Administration's Office of Advocacy focuses on additive manufacturing (also referred to as 3D printing) to determine if barriers, best practices, and big ideas found in that industry are instructive for technology industries generally. Moreover, small firms play an increasingly important role in U.S. manufacturing, and now account for almost half of America's manufacturing employment. Dense networks of these small manufacturers are vital to the process of taking a product from concept to market, and the exchange of manufacturing know-how across suppliers is essential for the diffusion of the new products and innovative processes that give U.S. manufacturing its cutting edge. However, these small firms face barriers to innovation, a key element in strengthening U.S. competitiveness. The book describes these barriers, and offers suggestions about ways to increase innovation by small manufacturers and improve the flow of innovation and information within supply chains. Furthermore, the accelerator phenomenon has been cited nationally and internationally as a key contributor to the rate of business startup success. Accelerators select and invite a small group of entrepreneurs to startup boot camps, providing mentoring, resources, and, most important, industry connections during these programs. This book aims to help entrepreneurs and policymakers by categorizing a variety of startup assistance programs to determine what factors distinguish accelerators from other programs, as well as create a starting point for developing meaningful metrics to determine the relevance of accelerators for policymakers.

In: Small Business Innovators
Editor: Mary Brown ISBN: 978-1-63483-265-6

Chapter 1

SMALL INNOVATIVE COMPANY GROWTH—BARRIERS, BEST PRACTICES AND BIG IDEAS: LESSONS FROM THE 3D PRINTING INDUSTRY*

Mark Harrison

FOREWORD

Innovation in the United States has been one of the driving forces in our development as one of the leading economies in the world. Innovations commercialized by U.S. companies have also benefited our society by allowing us to attain prosperity and a good quality of life. While large corporations and the federal government play important roles in the development of innovative products and services, the story is incomplete without the significant contributions and role of individual entrepreneurs and small, agile, high-growth businesses in developing innovative products in the fields of science, technology and engineering. Small companies comprise the overwhelming majority of all businesses in the United States and they must be able to effectively and efficiently bring their innovative products and services to market and grow. Therefore, as the United States moves into the future, we need policies and programs to support the development of an innovation

* This is an edited, reformatted and augmented version of a report issued by the U.S. Small Business Administration, Office of Advocacy, January 2015.

ecosystem that allows small innovative firms to grow, thrive, and create jobs—building the economy and a stronger America.

Implementing effective policies is more critical now than ever if the United States is to remain a leader in today's highly interconnected and hypercompetitive global economy. Both developed and developing countries have and are continuing to implement policies and programs to foster their own innovation ecosystems and economies.[1] However, innovation is not an easy or straightforward task despite the many resources dedicated to it. Small businesses trying to bring innovative technology products and services to market often face unique challenges and barriers not faced by large multinational corporations, and they must often overcome such barriers with fewer resources.

The Office of Advocacy's Innovation Initiative focuses on the needs and concerns of small innovative companies. The Innovation Initiative was designed to investigate the challenges individual entrepreneurs and small high-growth companies face in attempting to commercialize an innovative product or service. This report describes eleven key barriers to small innovative company growth. Each barrier is discussed in three parts:

- ***What We Heard***—Describing the input from entrepreneurs, private industry and other participants in the innovation ecosystem as a result of Advocacy's outreach efforts;
- ***Best Practices***—Identifying best practices or recent activities around an identified barrier; and
- ***Big Ideas and Recommendations***—Highlighting policies or programs that should be considered by the federal government to overcome such barriers.

Developing and exchanging new ideas on innovation is essential to the United States' ability to compete and to lead in the 21st century global economy. As President Obama stated in his 2014 State of the Union address:

> "We know that the nation that goes all-in on innovation today will own the global economy tomorrow. This is an edge America cannot surrender."

This report is one piece of this effort. We would like to acknowledge and thank the many individuals who shared their expertise and perspective on the subjects studied here. Special thanks go to the Institute of Electrical and

Electronics Engineers for sharing information on the results of its membership survey on challenges to technology company growth.

Winslow Sargeant, Ph.D.
Chief Counsel for Advocacy
Mark Harrison
Entrepreneur in Residence Office of Advocacy

January 2015

INTRODUCTION

The Office of Advocacy's Innovation Initiative

The Office of Advocacy launched the Innovation Initiative in 2013. The goal of this initiative is to identify the challenges and barriers hindering the growth and development of small innovative companies and offer solutions to surmount such barriers. The Office of Advocacy's Innovation Initiative is comprised of the following components:

- Outreach to innovation and entrepreneurial small business stakeholders,
- Communicating the feedback and concerns to appropriate federal agencies, and
- Reporting the initiative's findings.

Dr. Winslow Sargeant, chief counsel for advocacy, has led a coordinated outreach effort amongst his team, including 10 regional advocates, an Office of Interagency Affairs, an Office of Economic Research and, most recently, an entrepreneur-in-residence. Advocacy has sponsored a series of symposiums titled "Small Business and Government: Maximizing Entrepreneurship, Driving Innovation" aimed at bringing together individuals working in the innovation ecosystem to hear firsthand about the challenges to growing successful businesses. (See sidebar.)

Advocacy's outreach efforts over the past two years provided information and insight into emerging innovation sectors such as green technology, renewable/alternative energy, advanced manufacturing; much of this input falls into one of three categories, which Advocacy terms "the 3 Bs": Barriers,

Best Practices and Big Ideas. While Advocacy received feedback on many barriers and industry sectors, most of the concerns centered on the development of new technology by small businesses and advanced manufacturing. In FY 2013, Advocacy continued outreach efforts to the innovation ecosystem looked more closely at issues regarding new technology/product development and advanced manufacturing.

Advocacy focused on one particular innovation/technology industry in order to determine if barriers, best practices, and big ideas found in that industry are instructive for technology industries generally. The technology industry chosen was additive manufacturing (also referred to as 3D printing). The additive manufacturing industry met the criteria for further evaluation in that it is a subset of advanced manufacturing and is experiencing a fast pace of innovation.

The review of data and information on the additive manufacturing industry has been highly qualitative relying on the review of secondary source materials (e.g., research reports, white papers, articles, etc.) from organizations such as McKinsey, Gartner, PriceWaterhouseCoopers, Ernst & Young, universities, research laboratories, trade and business organizations, and the popular and trade press. Advocacy's entrepreneur-in-residence and regional advocates were in direct contact with more than 80 individuals in the additive manufacturing industry and the innovation ecosystem, including entrepreneurs, researchers, government officials, venture capitalists, and angel investors. Advocacy was also given access to data collected by the Institute of Electrical and Electronics Engineers, or IEEE, from its members via an online survey conducted in July 2014 on issues surrounding challenges to growth for technology companies.[2] This collective outreach was crucial to understanding the issues facing small businesses in the additive manufacturing industry and is one of Advocacy's core missions and strengths —providing insights from U.S. small businesses to policymakers on the issues that directly affect them.

Part One of this report focuses on various aspects of innovation as well as the additive manufacturing industry, which provides a framework for analyzing barriers to small innovative company growth. Part Two discusses the 11 barriers to small innovative company growth. For each, it presents relevant best practices or recent activities, as well as program and policy ideas and recommendations for consideration by the federal government.

Part One: An Innovation Case Study

Innovation

The federal government sees innovation as the key to future U.S. growth and international competitiveness.[3] According to a PriceWaterhouseCoopers survey, 80% of the participating executives identified growth as a top priority over the next five years and 93% believe that organic growth through innovation will be the primary driver of revenue growth.[4] An innovation is the implementation of a new or significantly improved product (good or service), or process, a new marketing method, or a new organizational method in business practices, workplace organization or external relations.[5] It is important to note that for new products and services to be considered innovations, they must be commercialized—introduced into the market.[6]

While there are numerous theories about the number of different types of innovation, we focused on small companies working on breakthrough or radical innovations[7] since these types of innovations present the highest potential to create jobs and new markets, and have substantial impact on the economy.[8]

How Many Small High Tech Firms Are There?

While small businesses make up approximately 99.7% of the 28.2 million businesses in the United States, only 5.7 million of these small businesses actually have employees. Based on internal calculations of Census data done for this report analyzing the number of firms in certain high technology industry NAICS codes, the Office of Advocacy estimates the number of small businesses in high technology sectors to be between 250,000 and 350,000.

Who Are the Innovators?

While the federal government and large corporations play a significant role in innovation in the United States, small businesses make unique and indispensable contributions as well. Media coverage of small high-tech companies developing new products suggests that there are a large number of such firms in the United States; however the Office of Advocacy estimates that around 250,000–350,000 small businesses are involved in high technology sectors.[9] Although relatively small in number, these small technology companies have demonstrated the ability for significant job creation.[10]

However, for these companies to create these jobs and other economic development benefits, they must be growing businesses and not just startups. Thus in order for the United States to obtain the maximum economic and societal benefits from innovation, policies must be in place that allow these small companies to start their firms, innovate, bring new products and services to market, and grow their businesses.

What Is an Innovation?

There are many definitions of "innovation," however the common thread is the commercialization of something new.

What distinguishes an innovation from an invention? Commercialization—or bringing a product or service to market.

Breakthrough innovations are new, bold and substantially superior to the next best available thing. This is what most people have in mind when thinking about innovation.

Radical innovations create drastic changes to the competitive environment for a product or service, or creates new businesses.

Sources: "The Measurement of Scientific and Technological Activities: Guidelines for Collecting and Interpreting Innovation Data: Oslo Manual, Third Edition," OECD, Paris. Capturing Change in Science, Technology and Innovation: Improving Indicators to Inform Policy, Robert Litan, Andrew W. Wyckoff, and Kaye Husbands Fealing, eds. "Four Types of Innovation and the Strategic Choices Each One Represents," Jake Neilson. Breakthrough Innovation and Growth, PriceWaterhouseCoopers.

Innovation and Manufacturing in the United States

As stated in the White House report, *Making in America: U.S. Manufacturing Entrepreneurship and Innovation*, the U.S. manufacturing sector is growing and is pivotal to the innovation economy. [11] The report states that the manufacturing sector employs 60% of private sector R&D employees in the U.S. and contributes 75% of private sector research and development. Manufacturing firms introduced a new product or service and new production or distribution method at twice the rate of non-manufacturing firms, and most high-technology manufacturing sectors innovate at twice the rate of all manufacturing firms.[12] Recent manufacturing innovations are providing an advantage to U.S. manufacturing firms by lowering production costs and reducing the time to get products to market. This has led to new manufacturing startups, increased investment in research and development, and increases in manufacturing employment and within related supply chains.[13]

One important manufacturing innovation creating advantages for U.S. manufacturers is additive manufacturing. As part of Advocacy's process of identifying barriers to small innovative company growth, outreach efforts have been focused on companies in the additive manufacturing industry. Reviewing barriers of companies in all technology industries can lead to overly broad conclusions given the differences in characteristics among emerging technologies. Choosing a specific industry enables a better understanding of the challenges small companies face in that industry which are likely to be applicable to the challenges in other technology industries as well.

The choice of additive manufacturing was driven by three factors:

- The technology is potentially a breakthrough or disruptive one with a significant opportunity for economic impact;
- The industry or technology is showing rapid advancement; and
- There exists or potentially could be significant small business participation.

The McKinsey Global Institute estimates that the economic impact of 3D printing could range from $230 billion to $550 billion by 2025 and noted that the rapidly improving technology is leading to an increase in its use.[14] Additive manufacturing is being used and has the potential to significantly affect many industries. Small businesses are participating in many areas of the industry (3D printer manufacturers, materials, print shops, etc.). In addition, the President has demonstrated strong support of advanced manufacturing in the United States including additive manufacturing. America Makes was founded in 2012 as a national accelerator for additive manufacturing and 3D printing. It was formerly called the National Additive Manufacturing Innovation Institute and is the pilot institute in President Obama's initiative to build a national network for manufacturing innovation. (America Makes is discussed later in Part Two of this report, in connection with barrier no. 2, R&D Support).

Additive Manufacturing

Additive manufacturing (AM) is increasingly being adopted in manufacturing and garnering significant attention in the media.[15] AM is defined as the process of joining materials to make objects from 3D models, usually layer upon layer, as opposed to subtractive methods.[16] While the technology has been around for almost 30 years, improvements in AM technologies are heading towards an inflection point. AM is currently used across diverse industries including those producing consumer, industrial, medical, automotive and aerospace products.

The AM process begins with a digital 3D model generated using computer aided design (CAD) software. The CAD file is then converted to an STL file which instructs the 3D printer to create or "print" the object applying material layer by layer. AM was first and is still primarily used for prototyping and modeling. As the quality of the technology and materials improved AM has been used to create tooling and finished parts. AM's largest potential is its increased use to produce finished parts/products.[17]

AM is a significant innovative technology development. Even if AM is viewed only as a technology for rapid prototyping, modeling and tooling, it is significantly affecting manufacturing by reducing the time and cost in the manufacturing process. However, as AM technical capabilities improve and it is increasingly used in the production of finished parts, the impact of AM will be even greater affecting product quality and functionality, business models, supply chains, manufacturing footprints, product logistics, etc.

The Additive Manufacturing Industry

The AM industry is growing. According to the Wohlers Report 2014, the compound annual growth rate for global revenues for all additive products and services during the last 25 years is 27%. In 2013, global revenues from all products and services were $3.07 billion which were almost evenly split between product revenue ($1.55 billion) and service revenue ($1.52 billion). There are wide ranging estimates of the size of the industry by 2020 from $21 billion to $230 billion; however today, AM is 0.03% of the $10.5 trillion global manufacturing market.

The market is global, with AM systems being produced and used in many countries located in Africa (South Africa, Egypt), Asia (China, India, and Japan), Europe (Germany, France, UK, and Russia), as well as in Brazil and Canada. However, the United States is the leader with 38% of installed industrial additive systems worldwide with Japan, Germany and China representing 9.5%, 9.1% and 8.8% respectively. Israel is the leader in production and sales of industrial additive systems with 54.7% of the market, followed by Europe at 21% and the United States at 18.6%.[18]

Additive Manufacturing or 3d Printing?

While the term "3D printing" has become synonymous with additive manufacturing, 3D printing is one of seven different processes for additive manufacturing recognized by the ASTM International Committee F-42 on Additive Manufacturing.

Source: The ASTM International Technical Committee F42 on Additive Manufacturing Fact Sheet.

The AM service industry is also growing. The typical AM service provider may specialize in one or two AM processes (each process requires a different machine) and can provide a range of services from production of prototypes to final parts. Two trends in the service industry are system manufacturers beginning to offer production services and an active M&A market for service providers.[19] In addition, there are new business models being created involving service providers and individual consumers. The development of 3D print shops where individuals can have an object 3D printed as a service is an opportunity for small businesses and is being piloted by large companies such as Office Depot, Staples and UPS. Another model is bringing together 3D designers, parts producers and consumers through the Internet. There are variations but basically 3D designers can post or sell their designs for an object (e.g., jewelry) for consumers to purchase and potentially print themselves or have the parts producer print the object and send it to the consumer. Companies in this space include Shapeways, Thingiverse, Cubify, and i.materialise.

Personal 3D Printers—The Maker Community

The growth in sales and use of personal 3D printers is evidence of a movement in the United States of people who actually want to make things themselves.[20] Personal 3D printers generally cost less than $5,000 and use only plastic as build material.[21] There are hundreds of manufacturers of 3D printers. One factor in the increasing growth in personal 3D printers is the expiration of the patent for fused deposition which is the process used by almost all personal 3D printers. These machines generally are not used for manufacturing or industrial purposes but mostly used by hobbyists and students.

However, as the quality of these printers improves they are increasingly being used for professional uses by small businesses. Similar to the improvement in quality of the cell/smart phone, these personal 3D printers will continue to decrease in cost but increase in capabilities and in quality, making AM more accessible to small businesses. Another difference between large industrial AM systems and personal 3D printers is that the personal printers tend to operate in an open source environment in terms of operating software, materials, etc. This can be attributable to the nature of the Maker movement. The Maker movement is generating a lot of interest. The White House hosted a Maker Faire in June 2014 that included 3D printing among the processes featured.[22] Despite not having a large impact on business applications, the Maker movement has given the public some idea of the potential capabilities of AM and thus has likely positively affected the pace of development of industrial printer systems.

Benefits of Additive Manufacturing

AM is increasingly being used for final product/part production. However the majority of people interviewed in connection with this report believe that AM is a complement to conventional manufacturing processes, and not a replacement for them.

Seven advantages of AM may result in lower total production expenses and faster times to market. These include:

- Design freedom: AM allows a designer to create complex shapes and complex parts that cannot be produced using conventional manufacturing processes. Designers are not constrained by the rules of design for conventional manufacturing processes. This results in advantages like part consolidation which results in less product assembly, less labor, and less production time.
- Eliminating tooling: The AM process requires no tooling thus eliminating a major expense in the manufacturing process and shortening the lead time needed to begin production.
- Shortened time between design and production: Manufacturers are able to iterate numerous designs allowing them to "fail quickly" and determine proper design for product. Once a product is designed, manufacturers are able to move to production much faster than with conventional production methods.
- Increased customization: AM allows for customization of product design because there is no need to retool the system to accommodate design changes. This makes design variations, whether simple or complex, essentially free. Manufacturers can customize products more

easily and less expensively than with traditional manufacturing methods and are able to offer customers more product variations.

- Reduced inventory requirements: AM allows a company to produce products on a just-intime basis which lessens inventory requirements.
- Distributed manufacturing: Final parts can be produced and distributed faster since the digital CAD file can be sent to an AM system anywhere in the world. Manufacturers could produce parts closer to their customers' locations reducing shipping expense and lead time.
- Altered structures: AM designs can often lead to complex structures that can perform the same function as a conventionally made part but with less weight. This process, called "lightweighting," is a major attraction for the aerospace industry. Lightweighting can be accomplished using topological optimization, which uses computational power to determine optimal product shape, size, and strength. In addition, AM manufactured parts, in many cases, can require less assembly. For example, NASA was able to reduce the number of parts of a rocket engine injector from 115 to two using AM.[23]

Challenges for Additive Manufacturing

Aside from barriers that are similar to others experienced by small innovative companies discussed in this report, there are several technical or industry specific challenges to the increased use of AM.

- Cost of AM systems and materials: Industrial AM systems, particularly systems producing metal parts, are expensive. Systems can cost from $50,000 to over $1 million. Materials can also be expensive on industrial systems costing from $100–$300 per kilogram for some plastics and $300–$1,000 per kilogram for metals.
- Slow build speed: AM systems build products at a slow speed. Depending on the part complexity, the process from production to final part could take over 24 hours. The build platform is also relatively small limiting the size of products produced.
- Repeatability: Parts produced using AM should be consistent since they are based on a digital design. However, the same design on different systems using different batches of the same material will not always produce an identical part. This is a larger issue when manufacturing metal products/parts compared to those made from

plastic. The AM processes for metal products involves the heating, binding and cooling of metals which may be altered for any number of reasons (e.g., laser power spiking that adds more heat to the metal during part of the build process) leading to lack of consistent part production. The industry is working to increase its knowledge of the thermal dynamics of building metal parts using different additive processes, materials, etc. In addition, the lack of tools to monitor the consistency of the part during the build process means defective products are not discovered until a post processing inspection.

- Lack of materials: Plastics and metals are the two primary materials used in AM. Some composites and ceramic materials are being used in certain applications, and this is a growing area of development. However, the lack of a range of build materials was consistently cited in interviews for this report. In addition, system manufacturers are requiring that companies use their proprietary materials with their systems. Failure to use the proprietary materials could void the system's warranty.

The impact of these challenges is that AM, when used to make final products, make products that are relatively small, high value, customized and complex. Solutions to these issues are rapidly being developed. As these issues become less of an issue for manufacturers, more products will be eligible to be made using AM due to decreasing production costs and production speeds, and increasing product quality and repeatability.

Moving Forward

AM is part of an industry trend towards digital manufacturing. In digital manufacturing computers are used to aid and/or control the manufacturing process. This is not a new trend—digital manufacturing has been used for decades. An example of digital manufacturing is computer numerical control (CNC) machines which use a computer to control machine tools to cut and/or shape material. Manufacturing is increasingly becoming an integrated process. Products move faster and more efficiently through the design to final product process, or from the "virtual to the physical," as manufacturing equipment and software, as well as design software get more powerful and sophisticated and as the use of automation increases.[24] Manufacturers are gaining the ability to design, model and test a product and the production process quickly and less expensively before a product is actually made. AM is important to digital manufacturing in that AM allows for rapid prototyping, modeling, etc. during the product development stage and for complexity and customization into final products without significant additional cost or time loss.

The next step in the development of digital manufacturing is its integration with the "Internet of things." Gartner, Inc., an IT research and advisory firm, defines the Internet of things (IoT) as "the network of physical objects that contain embedded technology to communicate and sense or interact with their internal state or the external environment."[25] For manufacturers, this means the development of smart factories. A manufacturer can connect devices (e.g., sensors, equipment, products in process, etc.) and network them over an Internet infrastructure, allowing the manufacturer to monitor, in real time, the activities and processes in the factory. [26] The manufacturing equipment will communicate with the products in process to ensure quality standards are met and that the production process is efficient. This is all monitored through computer software and will allow the manufacturer to take any required actions to improve efficiency to the extent that the networked system does not adjust itself.[27] The network could even be extended to gather information from the supply chain and customers. Cisco estimates that IoT will generate $19 trillion in value over the next decade.[28] A word of caution—as with any product or process becoming increasingly digitized and reliant on software, manufacturing companies will need to address issues around the storage, security and privacy of company data.

The movement towards digital manufacturing and AM will heighten some of the barriers to the growth and development of small innovative companies. For example, the need for more engineers and individuals with advanced manufacturing job production skills, the so-called skills gap, will grow with increasing amounts of digitalization and automation in manufacturing processes. As noted in a report drafted by the President and CEO of Siemens and the CEO of Siemens Industry USA, "to succeed in advanced manufacturing workplaces, workers need to possess the production skills to set up, monitor, and control the manufacturing processes, and the process design and development skills to continuously improve them."[29] The challenge for small companies to find, recruit and retain a skilled advanced manufacturing workforce will become even more difficult than it is currently.

The overall equipment costs and expenses for small businesses is another barrier that will be exacerbated due to the rise in digital manufacturing. As AM technology improves (e.g., faster production speeds, increased part size and better finish, more materials, etc.) costs will become lower and increase the range of products that can be made using AM. In addition, as the technology improves, the expense of AM systems should decrease potentially making small business entry into AM less challenging. However, these cost reductions will be accompanied by new ones, as small firms have to replace legacy equipment and upgrade infrastructure to fully participate in the digital manufacturing supply chain. In addition, the small companies will also need resources to address the data storage, security, and privacy issues attendant to digital manufacturing.

As digital manufacturing and AM technologies improve and industries develop, there will be new opportunities and casualties for U.S. small businesses. In the near future, small business will have opportunities as service bureaus, product designers, 3D print shops, online design/print firms, repair shops and retail (e.g., jewelry). Other higher end opportunities include the development of AM materials, products/parts, and medical devices. As the use of digital manufacturing and AM increases, there is likely to be significant disruptions within traditional manufacturing supply chains. Small businesses involved in storing products or inventory management (e.g., distributors, warehousemen, job shops, wholesalers), and those firms transporting products (e.g., shipping companies, import and export companies) will likely be adversely affected.

The Innovation Process

Understanding small business development and the innovation process is important when evaluating barriers to small innovative company business growth. Understanding the process will provide insights into the stages at which the barriers are occurring, possible reasons for the barriers, and potential solutions. While there are many ways in which small innovative companies grow their businesses, we have set forth a model innovative company development process with six stages, shown in Figure 1 and outlined as follows:[30]

- Discovery/Identified Market Need: The small business has an idea for an innovation. The idea could be based on one or several sources: new or existing research; previous development activities; the entrepreneur's determination of a market need; or customer feedback.[31]
- Technology Demonstration: The small business performs research on its idea. Activities include: experimenting, developing prototypes, testing and evaluating, assessing the market opportunity, etc.

- Product Development: The small business builds on the potential solution identified during the technology demonstration stage to develop a specific product or other innovation. Activities include: confirming research results, product design, pilot testing, establishing production processes, and developing go-to-market strategies. By this stage, the small business has likely filed for patent protection of its idea.
- Commercialization/Market entry: The small business has finalized its product and is ready to begin selling product into the market. There is an emphasis on sales and marketing as well as establishing certain business metrics such as cost effective manufacturing processes. The company is generating revenues between $0–$2 million.
- Early Stage Growth: The small business is growing generating up to $10 million in revenue and hiring additional employees.
- Economic Development Impact Growth: The small business is generating over $10 million in revenue, and has increased hiring employees, potentially made investments in land and buildings, has increased tax obligations and undertaken other actions that positively affect the economy (e.g., philanthropy).

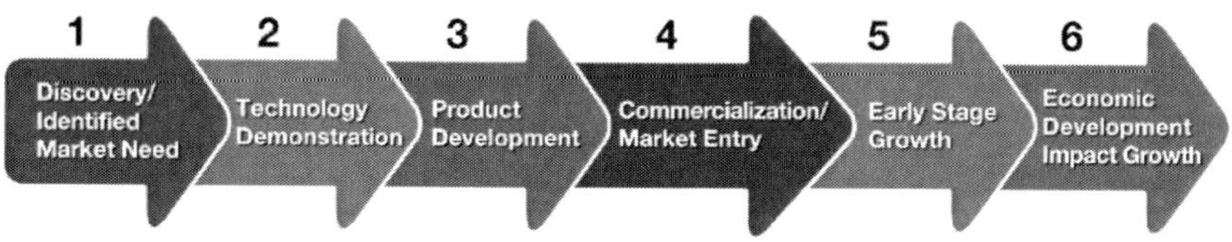

Figure 1. The Innovation Process.

There are a few things to note about this process:

- A company must reach the Economic Development Impact Growth stage in order to have the maximum impact on the economy and job creation. The economy will also benefit if the innovation creates a new or expands an existing market requiring other businesses to participate in such a market directly or in a supporting capacity.
- It is natural for the number of companies or ideas/products being developed to decrease as it moves through the process. All ideas don't necessarily get researched, all research does not get developed or commercialized and all companies don't grow to the point where they are having economic development impact.

- The process dynamic—it is rarely linear and is subject to numerous inputs. The entrepreneur will develop an innovation based on his/her knowledge base, existing external research, internal research, failed development attempts, existing products, market feedback, customer feedback, etc.[32] Entrepreneurs will move forward in the process, potentially stall (e.g., technical issue), will need to go back in process (e.g., more research or development activities), etc. This process can result in additional costs, the need for more or different resources and delayed commercialization/market entry.
- Delays in the process through commercialization/market entry can negatively affect companies attempting to develop new technology. Different technologies will move through the process at different speeds. Traditionally, products such as pharmaceuticals—one of the most heavily regulated industries—have long research, development and commercialization times (up to 15 years). Software, on the other hand, can be commercialized much faster (less than three years). Excessive delays in moving through the process for any reason (e.g., technical problems, lack of capital, legal or regulatory issues, etc.) can put a small company at risk for competitors with similar or competing technology moving through the process with greater efficiency to gain market share.
- There are many players involved in company and technology development who comprise our innovation ecosystem (which can be defined as individuals and entities that have influence and impact over innovation development and adoption).[33] Participants in the process include entrepreneurs, researchers, universities, federal laboratories, non-profit organizations, engineers, scientists, technical employees, small businesses, large corporations, angel investors, venture capital, crowdfunding providers, banks, incubators, accelerators, state and federal government as well as other support institutions and organizations.

The Role of the Federal Government in Innovation

The United States has many characteristics that make for a strong innovation ecosystem including the presence of world class universities, efficient capital markets for debt and equity, engineering talent, a large marketplace, and a strong legal system protecting intellectual property rights (IPR) as well as investor rights (e.g., bankruptcy laws, contract laws, etc.).

However, as previously noted, many countries are improving their innovation ecosystems meaning that the United States cannot afford to stand still. The federal government has taken steps to improve the innovation ecosystem, and many of these steps are highlighted in this report. However, the United States does have some growing problems which need to be addressed such as its workforce skill gap, reduced government spending on research and development, and an aging infrastructure (e.g., and the need to increase broadband accessibility, high speed transportation, etc.). In addition, efforts among participants are not always coordinated, leading to inefficiencies within the innovation process.

The federal government is a critical participant in the growth of innovative companies and the development and commercialization of new technologies. Many technologies used today can be traced back to research initiatives of or funding by the federal government including global positioning systems (GPS), microprocessors, the Internet, modern medicine, etc. In fact, the federal government, particularly the National Science Foundation (NSF), has played a pivotal role in the development of AM. The NSF has awarded almost 600 grants amounting to over $200 million for AM research and continues to play an important role in surmounting the technical challenges to AM.[34] The federal government's role is critical because its programs and policies generally affect all participants in the innovation ecosystem. Funding U.S. basic research, providing small business financial support through the SBA's loan guaranty and Small Business Investment Company programs, and purchasing small business products and services are just a few of the roles the federal government takes in support of small, innovation businesses.

Although the federal government plays a critical role in our innovation ecosystem, leadership of America's innovation process and ecosystem is dispersed among public and private sectors. Notwithstanding, the federal government should use its standing to make the innovation process work as open and efficiently as possible; this will help create a technical and business climate that provides innovative companies with opportunities to be successful. The government has many options to have an impact on and improve the U.S. innovation ecosystem including:

- Supporting funding for research and development, small business capital access, and education/job skills training either through direct financing or tax policy;

- Disseminating and providing access to government information;
- Organizing and facilitating cooperative interaction between participants in the innovation ecosystem;
- Supporting business opportunities and access to markets for small companies; and
- Enacting laws and regulations that support innovative company growth and technology commercialization.

The innovation process and ecosystem are fluid, and proposed changes to one part affects other aspects of the process or ecosystem. For example, access to capital has been a long standing problem for small innovative companies in growing their businesses. However, providing access to capital alone will not solve the problem. While no one would expect the government to move as quickly as inevitable changes in technology and the innovation process, it is important for government to be nimble and forward thinking when it comes to its innovation policies. Actual methods used by the federal government should be reviewed and changed as needed given the dynamics of the 21st century global economy.

PART TWO: BARRIERS, BEST PRACTICES, AND BIG IDEAS

The barriers discussed below are based on the feedback Advocacy received from companies, researchers, government agencies, universities, venture capitalists, lawyers, crowdfunders, consultants, trade groups, incubators, and other support organizations in the AM innovation ecosystem. This section identifies barriers and challenges specific to small companies attempting to commercialize a breakthrough technology product or service and grow their firms. The barriers identified are:

- The amount of student debt held by graduating students prevents them from pursuing entrepreneurial opportunities.
- The amount of funding and support of research and development in the United States needs to increase to ensure continued innovation.
- Entrepreneurs often lack information regarding market needs and product research and development efforts.
- There is a shortage of engineering and production job talent.
- Access to capital still remains a large barrier for small businesses growth.

- Small innovative companies have difficulty in commercializing products.
- Technology diffusion and adoption is harder for small businesses.
- The high costs of acquiring equipment and implementing a new technology is a barrier to entry for small businesses.
- Small companies need access to more business opportunities.
- Innovations often result in legal and/or regulatory challenges or uncertainty.
- Small companies continue to face challenges to exporting their products and services.

The barriers are in order of when they might be first encountered during the innovative company development process (See graphic below).[35] Many barriers affect companies in more than one stage of the innovative company development process with issues around access to talent and capital affecting all stages. While most of the barriers identified occur between the Discovery stage through the Commercialization/Market Entry stage, the importance of growing and scaling these small businesses for maximum economic impact cannot be overstated.

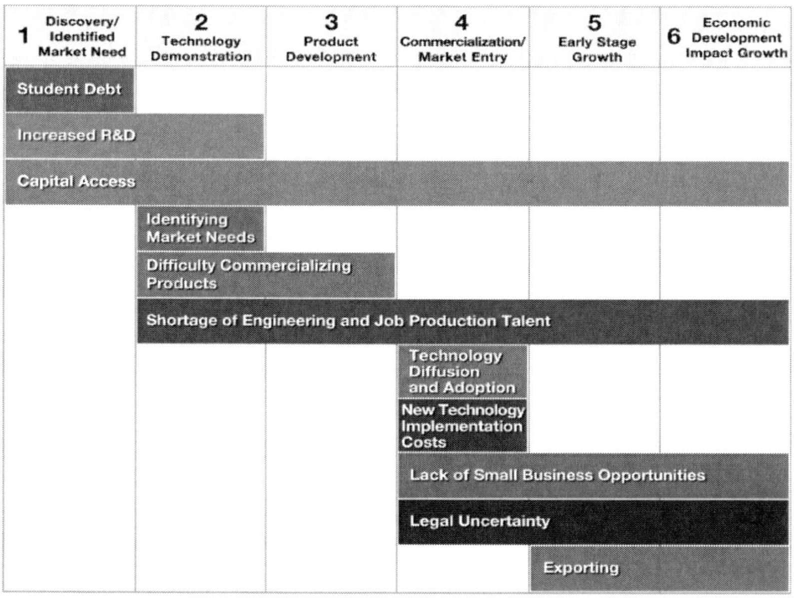

Figure 2. Barriers Along the Innovation Process Timeline.

Two other issues, immigration policy and infrastructure improvements, which are typically part of any discussion regarding challenges for growing small innovation firms, were not noted as major issues in the AM industry. These issues are not discussed in this report, since it is based solely on the feedback we received. However, implementing new policies and programs concerning immigration and improving U.S. infrastructure will only advance the goal of growing small innovative companies.

Each barrier is discussed in three parts:

- "What We Heard" from the people we interviewed and other background information,
- "Best Practices" regarding the barrier, and
- "Big Ideas and Recommendations," or potential proposed solutions to the barriers.

Barrier 1: The Amount of Student Debt Held by Graduating Students Prevents Them from pursuing Entrepreneurial Opportunities

1 Discovery/ Identified Market Need	2 Technology Demonstration	3 Product Development	4 Commercialization/ Market Entry	5 Early Stage Growth	6 Economic Development Impact Growth
Student Debt					

What we heard: Many college graduates are not pursuing entrepreneurial opportunities due to their outstanding student loans. College costs have been steadily rising over the last 10 years requiring many students to fund all or part of their education through student loans. Between 1995 and 2013, the share of families with education debt rose from 12 percent to 20 percent.[36] Total student debt outstanding for individuals under 30 years old in fourth quarter 2012 was $322 billion; this was distributed among 15 million borrowers for an average outstanding debt of $21,400.[37]

A person with significant outstanding student debt may forgo entrepreneurship for two reasons. First, the individual would be unable to meet their student loan payments because the entrepreneurial opportunity would not pay sufficient compensation. Second, the entrepreneur may have more difficulties raising money for their business due to the amount of their outstanding student debt (i.e., being over-leveraged). This may be a significant reason for the 23% drop in business starts by individuals aged 25 or under between 2002 and 2012.[38] There is a need for relief from student loans to allow talented graduates to pursue entrepreneurship.[39]

Individuals can receive student loan deferment or forbearance for reasons such as: enrollment in graduate school, active military service, unemployment or economic hardship (for up to 3 years), Peace Corps service, certain teaching positions, certain national service positions (e.g., AmeriCorps), etc. However, graduates wishing to pursue entrepreneurship are unable to defer their student loan obligations.

Best Practices

Income-Based Repayment Plan

President Obama has worked to improve the income based repayment (IBR) plan that allows student borrowers to cap their repayments for federal student loans to 10% of their discretionary income if they make their payments in a timely manner. The program lowers the amount of the monthly payment for those borrowers who have high levels of student debt relative to their income.[40]

Rhode Island Plan

The State of Rhode Island is considering a plan to defer or greatly reduce the monthly amount of student loan repayments for up to two years for those persons who launch or go to work for a startup in the state. If adopted, this appears to be the first program to use student loan deferment as a way to encourage entrepreneurship.[41]

Venture for America (VFA)

VFA is a two-year fellowship program that prepares recent college graduates for entrepreneurship by providing training and having the fellows work for startup companies. VFA will introduce its VFA Opportunity Award for Loan Assistance that will provide up to $5,000 to four fellows in the program to pay student loans during the first year of the program.[42]

Big Ideas–Recommendations

Entrepreneurship Student Loan Deferment

Similar to deferments to attend graduate school, the entrepreneurship student loan deferment would allow students to defer their student loan payments if they start or work for a qualified company. A qualified company would be any firm in business less than 4 years and has less than $5 million in revenue for each of its last 2 fiscal years. This program would be specifically targeted to allow students to pursue entrepreneurial opportunities without the immediate burden of repayment of student loans. Individuals starting or working for startup companies are likely to be working for little or no salary and need to commit as much of their resources as possible to building their business. While the IBR plan is a significant benefit for student borrowers, those pursuing entrepreneurship need special incentives for the risk they take.

Barrier 2: The Amount of Funding and Support of Research and Development in the United States Needs to Increase

1 Discovery/ Identified Market Need	2 Technology Demonstration	3 Product Development	4 Commercialization/ Market Entry	5 Early Stage Growth	6 Economic Development Impact Growth
Increased R&D					

What we heard: Research and development is a critical step in the innovation process. Research, particularly basic research, is just as valuable to the process as development. Basic research adds to the general body of knowledge about an area of study, usually with no regard for any potential commercial application. However, basic research has been the building block for subsequent breakthrough innovations. The federal government is the largest funder of basic research in the United States. Basic research is mostly performed by universities and colleges. Individuals working on or attempting to commercialize products based on basic research were concerned about a slowdown in federal research spending due to federal budget constraints. Everyone wanted increased federal spending but there was a split of opinion regarding whether the federal government should begin funding more applied research and commercialization activities and decreasing the amount of basic research it supports. It should also be noted that other benefits of research activities include: (1) training of skilled scientists and engineers and (2) the development of new processes or equipment to perform the required research.

Businesses and the federal government are the largest funders of R&D in the United States.[43] Businesses fund and perform most applied research and development activities.[44] Federal spending for basic research has remained relatively flat and while R&D by business has increased, policies should be implemented to encourage additional private sector spending on R&D. Such policies should enhance the ability of small businesses, often the sources of innovative products but financially and resource constrained, to be active participants in R&D activities.

U.S. support of R&D is important in light of such activities by our global competitors building their innovation economies. The United States has been the world leader in R&D activities, spending $424 billion in 2011[45] and forecast to be the leader through 2014 and to continue modest growth through 2020.[46] However, Asian countries, particularly China, are expected to increase their share of global R&D while the U.S. and European shares are expected to decrease. In fact, at their current respective growth rates, China is expected to pass the U.S. in R&D funding by 2022.[47]

The government should strongly support U.S. based R&D activities not only to maintain and enhance our ability to innovate and out of concern for the increased spending and growing capabilities of other countries, but for the potential economic impact and jobs created as a result of these activities.[48] While R&D is necessary for innovation, the government's and industry's future ability to support of R&D initiatives will depend, in part, on the state of our economy.

Best Practices

President's FY 2015 budget

The President understands the role federally funded research plays in innovation in the United States and has shown his commitment by increasing total federal R&D spending from $130.3 billion in FY 2013 to an estimated $133.7 billion in FY 2014 and is proposing an increase to $135.4 billion in his FY 2015 budget.[49] The President's proposed FY 2015 budget also includes a 1% decrease in spending on basic research, a 1.8% increase in applied research and a 2.3% increase in development activities by the federal government over estimated 2014 amounts.[50] The proposed FY 2015 budget also includes a provision to make the R&D tax credit permanent.

Global Innovation Initiative

Part of enhancing the quality of the research done in the United States is facilitating networking and cooperative work opportunities for U.S. researchers. An example of one such program is the Global Innovation Initiative.[51] The goals of this initiative are to build up the research talent pool and encourage collaboration between university researchers, faculty, students, and administrators in the United States, the UK, Brazil, China, India and

Indonesia. The administration should continue to foster different collaborative opportunities.

America Makes
America Makes was founded in 2012 as a national accelerator for additive manufacturing and 3D printing. Also called the National Additive Manufacturing Innovation Institute, it is the pilot institute in President Obama's initiative to build a national network for manufacturing innovation. It is based in Youngstown, Ohio, and has an extensive network of nearly 100 companies, nonprofit organizations, academic

America Makes

Based in Youngstown, Ohio, America Makes is an extensive network of nearly 100 companies, non-profit organizations, academic institutions and government agencies from all over the United States. America Makes was founded in August 2012 as the first National Network for Manufacturing Innovation (NNMI) Institute (see sidebar). It is focused on helping the United States develop and grow capabilities and strength in AM. America Makes supports and facilitates research collaboration among leaders from business, academia, non-profit organizations and government. Its research efforts are aimed at enabling technology transition from universities and government (i.e., basic research) through to commercialization. Its mission statement outlines five interrelated goals to increase the use of AM and help the U.S. manufacturing industry become more globally competitive:

- Fostering a highly collaborative infrastructure for the open exchange of AM information and research.
- Facilitating the development, evaluation, and deployment of efficient and flexible AM technologies.
- Engaging with educational institutions and companies to supply education and training in AM technologies to create an adaptive, leading workforce.
- Serving as a national institute with regional and national impact on AM capabilities.
- Linking and integrating U.S. companies with existing public, private or not-for-profit industrial and economic development resources, and

business incubators, with an emphasis on assisting small- and medium-sized enterprises and early-stage companies (startups).

The NNMI Institutes

President Obama has proposed building the National Network for Manufacturing Innovation (NNMI), consisting of regional hubs that will accelerate development and adoption of cutting-edge manufacturing technologies. In addition to America Makes, there are three other NNMIs:

Next Generation Power Electronics Manufacturing Innovation Institute, headquartered at North Carolina State University;

Digital Manufacturing and Design Innovation Institute in Chicago; and

Lightweight and Modern Metals Manufacturing Innovation Institute,Detroit Michigan.

Two additional institutes have been proposed: one focused on clean energy manufacturing for composites materials and structures led by the Department of Energy, and a biomanufacturing innovation institute led by the Department of Agriculture.

Big Ideas–Recommendations

R&D tax credit: We recommend the R&D credit be modified to allow more small businesses to take advantage of the credit by:

- making the credit permanent,
- simplifying the credit so that there are fewer disputes between the IRS and corporations using the credit,
- adjusting the amount of the credit to make the United States competitive with other world markets,
- allowing the amount of the credit to be higher for small companies and allowing small firms to deduct the credit against their payroll taxes, and
- making the credit transferable.

There seems to be some bipartisan support for positive changes to the R&D tax credit. The House of Representatives approved a bill in May 2014 making the R&D tax credit permanent and simpler to use.[52] The Senate Finance Committee approved a bill retroactively extending the R&D tax credit through 2015 and includes provisions for small businesses to offset the credit against their payroll taxes.[53] We make no recommendation as to the amount of the credit but it should be competitive with other countries taking into account the respective corporate tax rates and other tax incentives.[54] Allowing the credits to be transferable will allow small companies to improve their cash flow by selling the credits over and above their tax liability to firms that have income which can be offset by the credit.

A well-crafted R&D tax credit would help small companies participate in R&D activities.

Barrier 3: Entrepreneurs Often Lack Information Regarding Market Needs and Product Research and Development Efforts

1 Discovery/ Identified Market Need	2 Technology Demonstration	3 Product Development	4 Commercialization/ Market Entry	5 Early Stage Growth	6 Economic Development Impact Growth
	Identifying Market Needs				

What we heard: Many entrepreneurs in startup companies attempting to commercialize technology based on basic research often lack information on what products the market needs and wants. Many of these entrepreneurs are researchers commercializing technology that may be valid intellectual property (IP) but has no demand from customers in the market. Small business entrepreneurs that are commercializing technology based upon customer feedback or their experience have better insight into the market demand but, similar to the startup entrepreneur, often have no information about current or previous research or development efforts for a similar product. Both types of entrepreneurs are also often unaware of relevant existing patents that might affect their efforts despite the availability of databases with patent information.

Entrepreneurs could make more informed decisions about their commercialization activities by having access to this information. Imagine a startup entrepreneur learning early in the development process that there is no current demand for their technology in the market. Learning this information early allows the entrepreneur to make informed decisions about moving forward: Can I pivot and develop a product the market wants? Can my technology enhance another firm's technology providing a potential partnership opportunity? Do I stop my development efforts? This informed decision making makes the commercialization process more efficient because time and money are not spent on technologies that don't have a place in the current market.

Best Practices

National Science Foundation (NSF) Innovation Corps (I-Corps) Program: The NSF I-Corps Teams Program prepares scientists and engineers to focus beyond the laboratory by providing them access to resources to help determine the readiness to transition technology developed by previous or current NSF projects.[55] The program lasts about seven weeks and provides entrepreneurial training through a combination of guidance from experienced entrepreneurs, online curriculum, and group meetings.

The I-Corps Team program is open to teams that apply and are chosen to participate. Each team is comprised of a principal investigator, usually a professor who has done the research which is the basis for the I-Corps project, an entrepreneurial lead, usually a post doctoral or graduate student, and business mentor. The I-Corps curriculum provides real-world, hands-on immersive learning about what it takes to successfully transfer knowledge into products and processes that benefit society.[56] The goal is to get team members out of the university/laboratory setting and engage the market (i.e., address the market risk).

The NSF runs the program a few times a year in "cohorts," with each cohort accommodating 21-24 teams. Each participating team receives a $50,000 grant to fund their activities during the program. NSF has taken steps to expand the program by implementing five I-Corps Nodes as well as I-Corps Sites.[57] The NSF and National Institutes of Health (NIH) are collaborating on a pilot program to speed biomedical research technologies to market.[58]

Big Ideas–Recommendations

Expand NSF I-Corps Program: The I-Corps Program not only helps the innovation process by providing researchers with entrepreneurial training, but it also reduces investment risk by helping to identify those technologies with stronger commercial potential.

Based on conversations with individuals at the program, approximately 40% of I-Corps teams go on to start a company. Many pursue licensing opportunities, a handful of companies have received private funding and at least one has been acquired. Expanding this program to other agencies funding research can lead to an increase in technologies that ultimately are commercialized. While it is difficult to quantify a negative, the time and money saved as a result of technologies determined not to be commercially viable is another significant benefit.

Entrepreneur's Government Assistance Website: A website could be created where entrepreneurs can find information on technologies and tech industries that would assist them in deciding how to move forward with development efforts for their technology. The website would include information such as: current and recently funded research, current and soon to be expired patents, government generated market information, government financing resources and technology development programs and emerging technology briefings. (See discussion under Barrier Number 7, Technology Diffusion).

THE NSF I-CORPS

The National Science Foundation's Innovation Corps program (I-Corps) is a set of activities and programs that prepares scientists and engineers to expand their focus beyond the laboratory. The program aims to broaden the impact of select, NSF-funded, basic-research projects. I-Corps is an excellent example of a program bridging the distance between basic research, technology proofing, and product commercialization, a key mode of progress in the innovation ecosystem.

The goal is to create a one-stop location for this information presented in a way that it is assessable to and actionable by entrepreneurs. Having access to this information would assist entrepreneurs in assessing their proposed solution against other technologies being developed with the assistance of federal funding. This information would also assist entrepreneurs identify potential partners or other technologies that could help commercialize their product. Since all of the information is government generated, it is currently available to the public.

Thus, the raw data is available but needs to be presented in a manner so an entrepreneur can easily review and use the information. There are similar online initiatives such as: BusinessUSA.Gov, which is creating a centralized, one-stop platform to make it easy for businesses to access services to help them grow; the Federal Laboratory for Technology Transfer website, which allows individuals to search for expertise or technologies ready for licensing at federal labs; and SBIR.Gov, which contains information about the SBIR and STTR programs as well as open SBIR solicitations.[59]

Barrier 4: There Is a Shortage of Engineering and Production Job Talent

1 Discovery/ Identified Market Need	2 Technology Demonstration	3 Product Development	4 Commercialization/ Market Entry	5 Early Stage Growth	6 Economic Development Impact Growth
	Shortage of Engineering and Job Production Talent				

What we heard: There is a general consensus among technology firms that there is a shortage of engineering and production job talent which can slow company growth. The challenge is not only finding individuals with the requisite science, engineering (S&E) education and experience but also individuals with skills to work in production jobs (e.g., technicians, etc.). The mix of S&E and production job talent within any small company will vary depending on the particular technology industry and stage of company development. For small companies in the AM industry, the focus was more around the need for highly skilled production job workers than S&E workers. While the lack of individuals to fill these highly skilled positions will continue to be a problem, two important issues for small businesses in particular are: (i) how do they recruit and retain these skilled workers, and (ii) how do they get their employees access to the necessary training needed.

There are many efforts being made to increase student participation in science, technology, engineering and math which will help address the shortage of S&E workforce talent. Most small AM manufacturers we contacted indicated they were able to find individuals for scientific and engineering (S&E) jobs without much difficulty except for CAD engineers. The CAD engineer is the person actually using software to design the AM product. Geography has some impact in that firms located near technology hubs (e.g., Silicon Valley) are able to find these engineers easier than other areas of the country. However, the shortage of CAD engineers can lead to increasing competition for their services and subsequent salary inflation making it difficult for small companies to pay market salaries. Small firms' ability to recruit and retain engineering talent is becoming increasingly important.

As is likely the case in many new innovative technologies, including AM, small businesses may also need to retrain their current S&E workforce which may be challenging given their limited resources. Many individuals we spoke with agreed that design engineers must begin to think about creating designs that can take maximum advantage of AM processes. With conventional manufacturing processes, the design engineer must account for the requirements and capabilities of the manufacturing method and tools when designing a product. AM eliminates this concern, thus allowing the design of products with complex shapes and properties that can't be made using conventional manufacturing methods—a primary benefit of AM. Small business recruitment, retention and training of a skilled workforce is also an issue regarding production job workers.

Small businesses' ability to find individuals to fill skilled production jobs (e.g., technicians) often depends on the complexity of the parts or products being manufactured. The more complex the product (e.g., metal part for aerospace application) the higher the skill set required by the technician. Technicians may be required to check the CAD design, monitor the printer during build process, test product for spec compliance, and/or perform finishing work on a part. There are limited numbers of individuals with this skill set and many are expected to retire from the labor market over the next 10 years. Faced with a shortage of job production talent, some small manufacturers have turned to training their own employees. A number of small businesses we contacted that were training people for production jobs expressed a desire for a tax credit to offset all or a portion of their training expenses.

The talent gap challenge will loom even larger in the future due to the advent of digital manufacturing and rapidly changing manufacturing technology. As the technology improves, companies may find ways to streamline operations, increase automation, and/or change business models, all of which may affect the skill set required in manufacturing a product.[60] Further, as manufacturing software improves, individuals working in manufacturing will need additional training in computers and software as well as creative problem solving skills. Digital manufacturing and AM are growing; however the talent gap remains a major barrier to growth for many small businesses. It is critical that any training programs, apprenticeships or other solutions are providing individuals with the necessary skills to function in digital manufacturing.

Best Practices

Community college programs: There are many community colleges across the United States with technical training programs for AM and other jobs within advanced manufacturing and other innovation industries. Community colleges can often better tailor their training to the needs of a specific employer or industry. Individuals participating in these programs benefit by earning a portable credential and skills that allow them to earn a good salary.

NSF Engineering Research Centers:[61] The National Science Foundation's Engineering Research Centers (ERCs) are a group of centers located at universities across the United States that have partnered with one or more industry partners to work on various long range engineering challenges. An ERC provides an opportunity for industry to collaborate with faculty and students on innovations to address an identified challenge. The centers provide faculty and students with an excellent opportunity to gain experience working with business professionals that leads to constructive exchanges of ideas between academia and industry which benefits both.

Executive Action: In spring 2014, President Obama and Vice President Biden announced federal initiatives to support job training for individuals to help address the existing U.S. skills gap. The initiatives encourage business, community colleges, unions, state and local governments, non-profit companies and other training organizations to partner to develop "job-driven" i.e., training programs to prepare U.S. workers for the current and future job market. [62] While the initiatives are different, they have consistent goals so that the programs developed are: job-driven (i.e., responsive to the needs of employers); scalable, either national in scope or representing a best practice for other organizations; and providing participants with a career advancement path.

The three initiatives include:

- A job training competition administered through the Department of Labor. The competition's goal is to develop job training programs and it has received almost $500 million in funding;
- The American Apprenticeship Grants program which will award up to $100 million in funding to partnerships developing apprenticeship models; the program emphasizes models in high-growth sectors such as advanced manufacturing, information technology, technology services, and healthcare;[63] and
- The Registered Apprenticeship College Consortium which encourages agreements between colleges and registered apprenticeship programs to allow graduates of such programs to receive college credit for their apprenticeship training.[64]

The Workforce Innovation and Opportunity Act (WOIA): Signed into law July 22, 2014, by President Obama, the WOIA aligns certain federal programs and resources to assist job seekers gain employment as well as the education, training and support services required to succeed in the current and future job market, and help match skilled workers with employers. The Act was approved by a wide bipartisan majority and is the first legislative reform in 15 years of the public workforce system.[65]

Manufacturing Universities Act of 2014: Senator Christopher Coons (D.-Del.) introduced this bill to the Senate July 31, 2014. The legislation allows up to 25 universities to be designated as "manufacturing universities" and thus eligible to receive $5 million per year over four years to modify university engineering programs to emphasize advanced manufacturing. Manufacturing universities would have goals such as:

- increasing the number of students receiving undergraduate and graduate degrees in engineering or an applied science related to manufacturing,
- improving engineering curricula to focus on manufacturing,
- increasing the number of joint projects with private sector manufacturing firms, and
- increasing the number of students participating in internships. The manufacturing universities would also oversee interdisciplinary programs across its colleges, programs, and departments to advance manufacturing productivity and innovation.[66]

Big Ideas–Recommendations

Small Innovative company Apprenticeship Participation: The government should support and encourage the development of apprenticeship program models that facilitate Small Innovative company participation. Efforts are being made by various organizations to develop scalable apprenticeship models for advanced manufacturing.[67] Small businesses may face challenges participating in apprenticeship programs such as:

- lack of information about how to participate and the benefits of participating,
- lack of financial resources to fund apprentices, and
- too large an administrative burden to participate.

The government should support programs that include outreach strategies to small businesses as well as administrative requirements that are straightforward, may be done over the Internet and are not otherwise unduly burdensome. The government should consider providing a tax credit for small businesses participating in apprenticeship programs to incentivize their participation. The tax credit should be based on a percentage of compensation paid to the individual during the term of their apprenticeship and continue for up to two years should the individual be permanently hired by the small company providing the training. We believe the multi-year engagement between apprentices and small companies combined with a tax incentive will lead to increased and better recruitment, training and retention results for small firms.

Specialized Manufacturing Extension Partnership (MEP) Centers:[68] A select number of MEP centers should have increased capacity (e.g., equipment, personnel, etc.) for advanced/digital manufacturing processes and technologies (e.g., AM) to allow hands on training as well as access to equipment for small businesses. In the case of AM, a number of existing or newly created MEPs would specialize in AM. These MEPs would receive industrial AM systems capable of prototyping and parts manufacturing.[69] These specialized MEPs would provide a platform for both S&E and production job workers at small businesses to gain experience on AM equipment and processes. These centers could also serve as a facility where individuals getting training in AM from community colleges and technical schools could gain some hands-on experience. The government could defray some costs by seeking contributions of AM equipment, software and materials from AM system manufacturers.

Barrier 5: Access to Capital Still Impedes Small Business Growth

What we heard. Access to capital remains a major problem for small businesses. Access to capital was the top impediment to small business growth cited by 36.8% of the respondents in the IEEE Survey. It was also cited as one of the biggest challenges for bringing a product to market according to 44.8% of IEEE Survey participants. Small innovative companies are often able to access government funding to develop their product through the proof of concept and/or technology demonstration stage.[70] However, they face major challenges finding capital to move from proof of concept to completing product development and introducing the product to the market (commonly referred to as the "valley of death"). Institutional investors (e.g., venture capitalists) traditionally don't invest in companies during this period because of the risk that the product may not be commercialized and the resulting loss of their investment.

However, venture capital funds with less than $50 million in assets under management (micro VC funds) that invest in seed/early stage companies have been gaining in popularity over the last few years. Over 50% of venture capital deals that closed during the six-month period from November 2013 through April 2014 were made by funds with assets under management of less than $50 million, and 65% of them had less than $100 million.[71] While the growth of micro VC funds is a positive development, they tend to invest in Internet and mobile computing businesses rather than capital intensive industries.[72] Angel investors and crowdfunding have become increasingly important financing sources for small innovative companies.

Once companies have entered the market, they will need access to additional capital, both debt and equity, to grow their business. Angel investors and venture capital funds are active investors at this stage. It is still not easy to raise equity funding even though there is probably more capital available to small innovative companies at this time due to the advent of crowdfunding, and increasing angel investment and venture capital activity (including micro VC funds and corporate VC funds). Small innovative companies have challenges accessing bank financing because they often have little collateral besides their intellectual property (IP) to support their loan application and they may not have the financial performance required to qualify for a loan. The lack of collateral and lack of operating history or poor performance were the top reasons cited in the IEEE Survey why small innovative companies are unable to get funding.

Manufacturers of personal 3D printers have had some success with crowdfunding with at least seven companies raising over $750,000 on the crowdfunding site Kickstarter in the last two years. One of these companies, MD3 LLC, located in Bethesda, Maryland, raised $2.9 million in May 2014. In July 2014, Kickstarter had 274 crowdfunding campaigns involving 3D printing technologies. Seed and angel investment represented 64% of AM company financing.[73] Venture capitalist increased their investments in AM in 2013. As of 2014 Q1, there was $79 million invested in 15 deals representing growth in investment activity of almost 320%.

Best Practices

SBA loan process: The SBA has streamlined some of its underwriting requirements used by banks for small loans hoping to increase the number of SBA loans going to small businesses. Banks can now omit doing a cash flow analysis or reviewing the debt-service coverage on loans of $350,000 or less if the business owner satisfies the agency's other credit standards. The SBA hopes that by streamlining the process more banks will increase the number of small loans approved. While touted as a way to get loans to minority businesses, this change can also help small innovative companies, many of which have little collateral and weaker financial performance, gain access to capital as well.

Government venture investing: As part of the President's Startup America initiative, the SBA has developed and begun licensing early stage small business investment companies (SBICs) that focus on investing in high growth companies.[74] Early stage SBICs will invest at least 50% of their capital into early stage companies that are not yet profitable at the time of investment and the other 50% in small companies as defined by the SBA.

There are a number of states that have programs providing capital to early-stage innovative companies. Examples include the Maryland Venture Fund, the Innovate New York Fund, and the Michigan Pre-Seed Fund Loan Program.[75]

The Jumpstart Our Business Startups Act of 2012 (JOBS Act): The JOBS Act amended securities laws to relax the rules and expand the methods for small companies to raise capital. The JOBS Act included equity crowdfunding, a new Regulation D provision allowing the general solicitation of purchasers of company stock under certain conditions, and an increase in the Regulation A exemption from $5 million to $50 million (Regulation A+). There has been criticism of the JOBS Act, particularly regarding the crowdfunding provisions, that the proposed rules are still burdensome on small businesses trying to raise capital. In fact, final rules for crowdfunding and Regulation A+ have not been implemented as this report was being finalized and it is not clear how much the Regulation D general solicitation offerings are being used. There have been discussions and proposed bills for a JOBS Act 2.0. While the intention to facilitate capital formation for small business is good, implementation of this law is proving frustrating to the small business and investment community.

Big Ideas–Recommendations

Explore the Creation of an SBA IP Loan Guaranty Program: This program would guarantee loans to small innovative companies that have little to no collateral except for their IP. IP and other intangible assets are becoming more important to company growth and value. However, unlike physical assets, companies are rarely able to use their IP to secure a loan or investment potentially leading to increased capital costs. IP represents a growing source of capital (in the form of royalties and licensing fees) generating approximately $120 billion for all U.S. firms globally.[76] The problem is banks are uncomfortable lending against this asset class. Issues regarding valuation, obtaining a valid security interest and liquidation of the assets upon default are legitimate concerns by a potential lender.[77] To offset these risks, lenders may be extremely conservative underwriting such loans resulting in small companies receiving financing with a low loan-to-value ratio for their IP.[78] The SBA guaranty would allow the company to receive additional loan proceeds. The SBA should work with IP lenders, IP valuation experts, lawyers, etc. to determine if such a guaranty program could be implemented.

If the SBA decided to move forward with an IP guaranty program, it should convene regulators, bank industry groups, etc. to begin a campaign to educate the market about using IP and intangible assets to finance businesses. In an increasingly knowledge based market where IP will play an important part, U.S. companies need to be able to unlock the value of their IP to grow their businesses.

Support Angel Investors: Angel investors are a key source of funding for seed and early stage companies. In 2013, angel investors invested $24.8 billion into startup and seed stage companies (45% of total angel investment) with an average deal size of $350,000.[79] Policies should be adopted to incentivize angel investors to keep providing this critical source of capital to innovative companies. Recommended policies include:

- *Accredited investor definition:* Beginning in 2014, the Securities and Exchange Commission (SEC) must review the definition of accredited investor to determine if it needs to be modified for the protection of investors every 4 years pursuant to the Dodd-Frank Financial Reform Act. Any increase in financial thresholds in the current definition will lead to a decrease in the number angel investors.[80] According to the SEC, an increase in the net worth requirement from $1 million to $2.5 million would decrease the number of eligible accredited investors by

60%.[81] The Angel Capital Association has indicated that 25% of their 12,000-plus members would become ineligible. The loss of this many angel investors will be highly detrimental to small innovative company fundraising. It is recommended that the SEC be required to consider the potential adverse impact on the number of angel investors should it change the financial thresholds in the definition of accredited investor. No change should be adopted by the SEC if it will result in a potential decrease of more than 10% of eligible households that would qualify as accredited investors under the current definition. If the SEC does change the definition causing a 10% decrease or more, it should then consider alternative methods for individuals to qualify as accredited investors such as relying on an investment professional, possession of a professional certification, etc.

- *Angel investor tax credit:* The U.S. should enact a national tax credit for investments by angels into small companies working in high technology fields. Investors would receive a 20% credit if they hold the equity for a minimum of three years with the credit being allocated as follows: year 1—0%, year 2—10%, and year 3—10%. Investors in companies working on designated technologies that take longer to develop (LD technologies) such as biotechnology and pharmaceuticals would receive a 40% credit if the equity is held a minimum of five years with 10% of the credit being allocated in years 2 through 5. The higher credit should incentivize investors to invest in LD technologies which need higher amounts of funding to get to market. In addition, given the issues of access to capital for minority and women entrepreneurs, the amount of the credit would increase by 5% in the final year for angels investing in minority- or woman-owned technology company.[82] To increase the attractiveness of the tax credits, they should be tradable and not be eliminated by the alternative minimum tax. The American Opportunity Act of 2011 that was introduced in the Senate (S. 256) contained an angel tax credit proposal that never became law but could be a starting point for drafting legislation.

Technology Commercialization Assistance Fund: This fund would add to existing commercialization assistance tools, such as SBIR Phase III.[83] This new fund would invest in companies that have successfully completed a SBIR Phase II project that needs additional financing for commercialization activities. Limiting potential investment applicants to successful SBIR candidates allows the government to vet the technology and management team before committing to investing additional funds. The investment would be a royalty-backed security, secured by the company's IP. The government would receive total payments of between 125% and 200% of the investment amount over a repayment period of 10 to 12 years. The amount of the repayment would increase the longer the company takes to repay. This incentivizes the company to repay the investment sooner. The cap on the repayment amount should also be attractive to current or potential equity investors because they know the fixed repayment amount to the government and do not have to share in the upside of the potential increased equity value in the future. By requiring repayment, the government can use the investment proceeds to help fund future commercialization efforts. The fund does not conflict, in letter or spirit, with the Bayh–Dole Act because the fund is specifically funding commercialization activities and not research. Fifty-four percent of respondents in the IEEE survey believed it would be fair if the government receives a return on investment for funding a company's commercialization activities.

Improve and re-implement SBA Participating Securities: The SBA licenses and regulates SBICs that are privately owned and managed investment funds that use their own capital plus additional funding from the SBA in the form of SBA guaranteed debentures (known as "leverage") to make equity and debt investments in small businesses. SBICs are eligible to receive two dollars in leverage (SBA debentures) for every dollar of private capital raised. The SBA debentures generally have a maturity of 10 years with semi-annual interest payments.

In 1992, the SBA created a new form of leverage known as a "participating security" which essentially gave the SBA a preferred limited partnership interest in the SBIC. The SBA was entitled to a preferred return and a share of the profits but did not receive regular periodic payments unlike the interest payments current SBICs make using the debenture leverage.[84] The participating security leverage was discontinued in 2004 after the technology bubble of 2002.

Re-implementing the participating security leverage would make it easier for SBIC funds to provide equity capital to growing small businesses. SBIC fund managers would be able invest in the equity of small businesses, which typically does not have a regular payment requirement, without having to be concerned about making an interest payment to the SBA (as required by the debenture leverage). This would be the same as standard venture/growth capital equity investments. Adjustments should be made to the terms of the participating securities to reduce the risk of loss to the SBA (e.g., only fund managers with early stage equity investment experience should be eligible for participating securities). Bringing back the participating security may lead to an increase in experienced managers in the SBIC program and an increase in available sources of equity capital for small growing innovative companies.

Barrier 6: Small Innovative Companies Have Difficulty Commercializing Products

1 Discovery/ Identified Market Need	2 Technology Demonstration	3 Product Development	4 Commercialization/ Market Entry	5 Early Stage Growth	6 Economic Development Impact Growth
	Difficulty Commercializing Products				

What we heard. Many small companies have difficulty commercializing (i.e., begin to sell to market) their technology products. The primary reasons challenges identified include these four:

- Management inexperience: During the development period before commercialization, the entrepreneur/management is not only finalizing technical aspects of their product but also taking steps to introduce and begin selling it in the market. Many researchers and small entrepreneurs may have experience building products but not necessarily taking a product to market.
- Lack of resources: The resource cited as most lacking by entrepreneurs was capital but other resources such as expertise, testing equipment, product designers, etc. can also lead to problems bringing a product to market.
- Lack of time: Many small companies already selling products into the market are challenged to finding the time and development resources to work on commercializing new products because they are so focused on growing their existing lines of business.
- Incomplete product or no market: Many products based on research ideas fail to reach the market because they are only part of a full commercial product or solution or there is no current market for the product. However, these "failed" products may be picked up by another researcher or entrepreneur in the future when the product can be a critical part of a new commercial product.

Many small businesses also expressed reservations about partnering with large companies to develop their technologies. While acknowledging the many benefits of working with large companies such as access to personnel, equipment, financial resources, etc., entrepreneurs were concerned about protecting their IP or being able to negotiate an equitable split on any IP developed as a result of the collaboration.

Best Practices

NSF I-Corps Program: This program, discussed under Barrier Number 3, would also help strengthen companies' commercialization efforts.

Collaborative work environments: These include accelerators, clusters, manufacturing institutes, andMEPs. Entrepreneurs consistently state that being able to brainstorm or collaborate with colleagues is extremely helpful in solving technical and business problems.

Big Ideas–Recommendations

Expand the NSF I-Corps Program: See discussion under Barrier Number 3

Commercialization training program: A program should be developed, similar in structure to the NSF I-Corps Program, to teach small firm management teams about the product commercialization process. The program would be available for teams that have been awarded a SBIR Phase II grant. Focusing teams on commercialization planning early during the product development stage will lead to better odds of the product reaching the market. There are examples of agencies providing some commercialization training, providing an opportunity for collaborative efforts and sharing of best practices. These include the Department of Defense (DARPA Small Business Planning Tool), Department of Energy (Commercialization Assistance Program) and NIH (Niche Assessment Program).[85] As part of the program, a database should be developed of CEOs who have successfully commercialized a product through the SBIR process and who are willing to serve as mentors or consultants to teams to assist them commercialize their product.

Commercialization assessment: A commercialization assessment should be performed on technologies after the first year of the SBIR Phase II grant. The assessment would evaluate the strength of the team's technology or product against what is currently in the market, the strength of its IP position, and potential market barriers and opportunities. As a result of the assessment, teams can improve their commercialization plans as well as determine how to best allocate resources going forward. The assessment may also save time and money for the team and the sponsoring agency if there appears to be little likelihood of commercialization. The committee performing the assessments would consist of agency representatives as well as outside professionals experienced in the technology and product commercialization. All SBIR Phase II awardees would be required to go through an assessment except those that either have received funding from and institutional investor or have established an affiliation with a large corporation.

Barrier 7: Technology Diffusion and Adoption Is More Difficult for Small Businesses

1 Discovery/ Identified Market Need	2 Technology Demonstration	3 Product Development	4 Commercialization/ Market Entry	5 Early Stage Growth	6 Economic Development Impact Growth
			Technology Diffusion and Adoption		

What we heard. Technology diffusion is the process of building awareness and distributing technical information and know-how about a technology for later adoption and use in the market by customers. Small businesses are often challenged learning about a new technology.[86] In the case of AM, many small companies have read or heard about AM but do not understand the technology's capabilities and are not clear about the benefits to the business.[87] Some of the barriers to AM diffusion and adoption by small businesses include:

- Technology evaluation: Small firms may not have the in-house personnel to evaluate AM technology and potential applications.
- Lack of time: Small firm management may not have sufficient time to evaluate AM due to focus on current business.
- Business case: Small firms may not have access to information or the proper in-house framework to evaluate the business case for adopting AM (e.g., cost-benefit analysis, return-on-investment).
- Cost: Small firms may not have a sufficient budget to experiment and evaluate a new technology particularly if such activities require a large financial commitment as is the case with AM. There is a fear of wasting time and money if the technology doesn't work for their business.

- Workforce skills gap: Small firms may not have the personnel with proper skill sets to adopt AM.
- Reluctance to change: Companies may have already invested in assets and infrastructure supporting conventional manufacturing processes and will not want to abandon those lines of business. Further, a new technology like AM may affect the company's business model and its relationship with suppliers and customers. Management may not want to pursue AM due to these potential disruptions.

Entrepreneurs who lack access to information about the capabilities of any technology, including AM, and its benefits are then faced with the challenge of determining which external sources of information to believe and trust or who can help them.

The challenge of technology diffusion and adoption is clearly a barrier for those companies considering incorporating a new technology into its business. However, it is also a barrier for small firms that commercialize new technologies because if the technology is not adopted by the market, the commercializing firms will be unable to grow.

Like many other instances when new technologies come to market, another barrier for companies adopting AM has been the lack of standards regarding AM materials and processes. AM processes involve building an item layer by layer by binding materials (e.g., metal, plastics) using heat. Researchers and industry are trying to determine the impact of the thermal dynamic processes and properties and attributes of products made from different materials, particularly of metals, that are subject to various AM processes. For example, does a metal part made using one AM process have different characteristics from the same part using a different AM process? Is one part stronger? More malleable? Do they react to temperature changes differently? AM parts and processes must have known and predicable qualities and be repeatable in order for AM to gain widespread acceptance. This is especially important in any industry (e.g., aviation) where the manufacturing process is required to be certified (in addition to the parts themselves). Standards to test for these characteristics also need to be developed. The industry is well aware of this issue and standards are currently being developed.

In 2009, the ASTM International Technical Committee F42 on Additive Manufacturing was established for "the promotion of knowledge, stimulation of research and implementation of technology through the development of standards for additive manufacturing technologies."[88] The ASTM F42

Committee has partnered with the ISO Technical Committee 261 on Additive Manufacturing to develop standards. The ASTM F42 Committee has over 200 members from 16 countries; it has approved 10 AM standards and another 20 are being developed. The standards are available for consideration by all global markets. Their development will help increase the adoption rate of AM and move the industry forward by providing a common set of rules, definitions, processes, and guidelines which will improve product quality and safety. The development of standards and cataloging of material properties of the various AM processes will also help identify which ones are best suited for a particular application. Standard-setting will help small companies make the informed choices regarding equipment and materials to use for their business.

Best Practices

The National Institute of Standards and Technology's Hollings Manufacturing Extension Partnership (MEP): MEP is a national network of centers located in the 50 states and Puerto Rico that work with small and mid-sized U.S. manufacturers to help them increase their efficiency by providing services focused on productivity improvement, product and workforce development, business practices and technology transfer.[89] The MEP's technology scouting program works with small and medium-size manufacturers to help find technology solutions to product or production problems.[90] For this program to be more effective for manufacturers interested in AM, more MEP facilities would need to add equipment and expertise for both AM prototyping and product production. (Also see the discussion of specialized MEP centers under Barrier Number 4, shortage of engineering talent.)

America Makes: See the discussion of America Makes, under Barrier 2, increasing R&D support.

Big Ideas–Recommendations

Small business technology diffusion strategy: The government should develop a strategy specifically focused on the diffusion of manufacturing technologies to small businesses. Many of the barriers to diffusion involve lack of access to information. The government is positioned to have access to much information about technical capabilities and business metrics surrounding new technologies. What information it does not have, it can certainly develop itself or in partnership with the private sector. To effectively disseminate the information, one or more "trusted sources" must be created. In the case of AM, America Makes is a natural choice. All research information about AM that has been developed or sponsored by the government should be housed at America Makes or other trusted sources, and it should be made available in a searchable database accessible to small businesses. The information can include technical capabilities as well as impact studies of AM on return-oninvestment, business models, energy consumption, and other key concerns.

In addition to trusted sources, the strategy must include getting the information out to small businesses. The government and trusted sources should develop specific content for small businesses and leverage existing networks like MEP centers to disseminate the information. Finally, adoption rates should be continuously monitored to determine the effectiveness of the strategy.

Emerging technology briefings: The government or other institution, potentially in partnership with the federal government, should take the lead in publishing a briefing directed at small businesses on technical capabilities, business metrics, and applicable laws and regulations specific to emerging technologies key to the U.S. economy. These briefings would provide small businesses with concise information on a new technology such as current and potential applications, market information, cost models, as well as the legal and regulatory landscape for those developing or adopting the technology. This information would provide an excellent starting point for small businesses trying to decide whether or not to pursue the development or adoption of a particular technology. The briefings would be available on the websites of the appropriate government agencies and the trusted sources.

Coordination among NMIIs: The technologies being studied at the various NMIIs cut across many applications thus providing opportunities for technological interaction. The diffusion of all the technologies involved would be enhanced with some coordinated effort among the NMIIs. For example, America Makes should coordinate efforts with the Digital Manufacturing and Design Innovation Institute and the Lightweight and Modern Metals Manufacturing Innovation Institute, both of which started operations in February 2014. Increased coordination between the NMIIs could also lead to focused research initiatives that may enhance the technologies being studied.

Barrier 8: High Equipment Costs Are a Barrier to Entry for Small Businesses Implementing New Technologies

1 Discovery/ Identified Market Need	2 Technology Demonstration	3 Product Development	4 Commercialization/ Market Entry	5 Early Stage Growth	6 Economic Development Impact Growth
			New Technology Implementation Costs		

What we heard: In some technology sectors (e.g., software), equipment costs have decreased significantly making it easier for small businesses to enter the market. However, in advanced manufacturing and AM, equipment cost can still be a barrier to technology implementation and growth. The cost of industrial AM systems is increasingly making it difficult for small businesses to enter the market. The average selling price for an industrial system was $90,370.[91] Prices of high-end systems capable of producing metal parts can be over $1 million. In addition, material prices and system maintenance costs also drive up the expense of using AM. Over time, the costs of industrial systems, along with material and maintenance costs, are expected to decrease even as the capabilities of these systems improve.

Many small businesses are using lower-end industrial systems or personal 3D printers to experiment with the technology. This level of capability is suitable for some businesses.. There are companies establishing networks of 3D printers around the world. [92] Customers are able to upload an STL file, receive quotes and choose among geographically dispersed networked members that own a 3D printer to make the object. The networks appear to be comprised mostly of individuals and small firms using 3D printers, not large industrial AM systems; they are being used for prototyping, modeling and do-it-yourself or hobby projects. The networked printer model can help some small companies enter the AM market and would become more powerful as the capabilities of the 3D printers improve. However, for those small businesses trying to enter the market for finished AM parts, the expense associated with AM technology will remain a potential barrier. In addition, small firms will also incur costs to make their equipment and processes compatible with digital manufacturing requirements including issues surrounding data storage, security, and privacy.

Best Practices

None identified.

Big Ideas–Recommendations

Innovation Equipment Lease Guaranty: The SBA should create a program whereby it would guaranty certain equipment leases by small businesses. The small business must be leasing new equipment that is an upgrade to its current equipment. The lease must involve total equipment cost of at least $50,000. An upgrade would include equipment with new capabilities and functionality enabling the production of higher quality products. It would not include replacing older model equipment with the newest version. For example, a small manufacturer that was either new to AM or had been experimenting with personal 3D printers could use the program if it was adding a new lower-end industrial AM system. That same business could also qualify if it then wanted to lease a higher-end industrial system that used metal materials.

Barrier 9: Small Companies Need Access to More Business Opportunities

1 Discovery/ Identified Market Need	2 Technology Demonstration	3 Product Development	4 Commercialization/ Market Entry	5 Early Stage Growth	6 Economic Development Impact Growth
			Lack of Small Business Opportunities		

What we heard: Small companies want the government to make a stronger effort to buy more goods and services from them. Many small innovative companies believe that government can do more to support small firms selling innovative products and services. Government procurement can involve the purchase of existing products and services, pre-commercialized products and services requiring some level of R&D (e.g., Phase III SBIR), and newly created products and services that did not previously exist.[93] Government procurement of innovative technology products and services can help develop the market for such products and services. Government use and acceptance of the product or service can also help lower the perceived risk of adoption by the private sector.

For small businesses, the challenge most often cited was the complexity and administrative burden of government contracting. This included businesses selling existing products and services as well as those selling pre-commercialized products. Many small firms noted that the amount of time, money and effort required to respond to government contract opportunities is simply unfeasible, particularly as there is no guarantee the firm will get a contract.[94]

For its part, the government has similar challenges to adopting an innovative product as the private sector. These include technical risk such as whether the product will perform as required or even be fully developed, and organizational risk such as aversion to trying a new technology, compatibility issues with current equipment or processes, and costs associated with new product fitting within budget.[95] Government must also balance the need to help small innovative companies through procurement with the need to wisely spend taxpayer dollars and make smart purchases to help government provide quality services to U.S. citizens.

Best Practices

RFP–EZ: RFP-EZ was a pilot program designed to make it easier for small innovative companies, which historically have not done business with the government, to find and submit bids or proposals on federal contract opportunities less than $150,000.[96] The goals of RFP-EZ were to reduce administrative costs, improve opportunities for small businesses, increase contracting efficiency, and decrease the burden on small business and contracting agencies. It would be worthwhile to evaluate how effective this pilot program was, in particular in decreasing contracting time and increasing contracts to small technology companies.

Challenge.gov: Challenge.gov contains a list of challenges and prize competitions run by more than 50 agencies across the federal government.[97] Government agencies solicit innovative solutions from the public to address mission focused technical and scientific problems. It essentially is the government procuring products/solutions that don't currently exist. Since 2010 federal agencies have held over 280 challenge and prize competitions. A recent review of the website showed challenges ranging from zero (i.e., nonmonetary prizes) to $15 million. These competitions provide an opportunity for small innovative companies to submit their solutions and potentially begin selling to the federal government.

Big Ideas–Recommendations

Follow up and strengthen RFP-EZ and Challenge.gov: Since the RFP-EZ pilot has ended, follow-up evaluation is needed to assess its effectiveness. We recommend increased use of a tool such as RFPEZ by federal agencies and an increase in the contract limit to $2 million. An increase in the complexity of the tasks and award sizes would increase the likelihood of participation by small businesses. The amount of the award needs to be enough to entice an entrepreneur or small company to submit a proposal. In addition, websites like

Challenge.gov, focused on opportunities for small innovative companies, will benefit these businesses.

Create a small innovative company database: A database of small innovative companies should be created containing comprehensive profiles relevant to government contracting for each firm. These profiles should be set up to allow a company's information to populate all contracting forms from various government agencies should the firm decide to pursue a specific contract opportunity. The ability of small firms to complete various agency contracting forms from its profile would substantially decrease the time and effort required to pursue government contracts. The database would also allow government agencies to search for firms with particular capabilities and invite them to bid on contract opportunities instead of only relying on firms that respond to a request for proposal.

Expand use of agency "other transaction" authority: Other transaction (OT) authority allows certain agencies to buy innovative products from small businesses. This tool should be expanded to other agencies.

Certain federal agencies use OTs to obtain or advance R&D or prototypes. An OT is not a contract but a special vehicle that allows the government the flexibility to get access to leading edge R&D and prototypes outside the requirements of the federal procurement regulations.[98] This OT authority is meant to attract companies that are not traditional contractors or do not comply with the federal procurement regulations.[99] The following agencies have OT authority:[100]

- Department of Energy
- Department of Transportation,
- Department of Defense,
- Department of Homeland Security,
- Federal Aviation Administration,
- NASA,
- National Institutes of Health, and
- Transportation Security Administration.

The agencies with OT authority could use it now with small companies for R&D and prototypes.

In brief, we are suggesting two things: expanding OT authority for any product or service from a small innovative company, and giving additional federal agencies permission to use OTs to purchase from small companies. Small innovative companies would benefit greatly by simply expanding this existing vehicle for providing products and services to the government outside of the normal federal contracting procedures.

Barrier 10: Technology Innovations Often Result in Regulatory Uncertainty and Legal Challenges

1 Discovery/ Identified Market Need	2 Technology Demonstration	3 Product Development	4 Commercialization/ Market Entry	5 Early Stage Growth	6 Economic Development Impact Growth
			Legal Uncertainty		

What we heard: There was strong agreement among the entrepreneurs and organizations contacted that the U.S. legal system strongly protects IP rights and contract rights which is important to innovation. IP was very important to 32% of small businesses participating in the IEEE Survey. Companies that are developing personal 3D printers and/or AM industrial systems or using them in their operations are not currently facing any significant legal or regulatory issues or burdens. Small companies we contacted indicated the only regulatory issues involved those around worker safety and the environment which they believed was adequately addressed by the current regulations. This is likely due in part to the relatively small size of the industry at this time. As AM grows, the level of scrutiny on the industry, its participants, the technical processes and various product applications will increase leading to additional laws and regulations being passed. In this regard, AM is like most new technologies with innovations around a new technology preceding regulation of its uses. While different technologies will trigger different legal issues, there were a few issues regarding AM that could limit small company growth due to legal and/or regulatory uncertainty.

Patents: The primary concern expressed among small innovative companies regarding patents was that the expense to patent a product is burdensome. small innovative companies incur patent costs prior to generating any revenue from the product, thus requiring the companies to find a source of funding for the patent expenses. Although not technically a legal issue or uncertainty, high patent expense is mentioned here because almost every entrepreneur we contacted cited this as a barrier despite the USPTO requiring lower patent fees for small and micro entities.

Patent infringement lawsuits among AM manufacturers have increased in the last few years.[101] The potential threat of a patent lawsuit may have a chilling effect on small innovative companies attempting to commercialize a product; an actual lawsuit against a small innovative company could impair the firm's ability to survive and grow. However, to date, patent infringement entities or patent trolls have not appeared to cause significant problems in the AM industry, but they have become a problem in many other technology sectors.

Another issue is the impact of AM on IP rights and protection. An exhaustive discussion on the range of issues on this subject is beyond the scope of this report.[102] However, one of the core issues is how do you protect the rights of patent and copyright owners when AM technology provides a means for unfettered infringement? How this issue is resolved, either through court cases or legislation, will significantly affect the AM industry and its small business participants.

Liability: Liability for injury caused by an additive manufactured product is still somewhat an open legal question. The question of whether the CAD file designer, the AM system manufacturer, or the product producer, etc. is liable for an injury is likely to be resolved in due course via the legal system. The issue is less clear when the producer of the product is not in the business of manufacturing which is becoming a likely scenario as the personal 3D printer industry continues to grow.[103]

Stifling Technology Development: For all of AM's positive applications, it also has the ability to help individuals commit bad or illegal acts such as counterfeiting, weapon production, etc.[104] Policymakers and regulators should bear in mind that people have been doing bad or illegal acts for a long time, and AM is just another technology that can be used to facilitate such acts. As they consider the proper policies to enact, policymakers must balance society's need for protection against bad acts against the societal benefits from the technology's full development. It's important that policymakers try not to put in place measures that can stifle the development of the technology. For example, instead of enacting policies limiting 3D printer functionality to render them incapable of printing certain items, consider requiring methods, such as steganography technologies, that will allow the attribution or tracking of products back to the 3D printer that created it.[105] It is also important that any policies enacted to counteract bad acts not place burdensome administrative requirements or monetary obligations on small firms trying to develop AM or other technologies.

Limiting Market Access: Imposing export controls on certain technologies can impair a small innovative company's growth potential. Export controls limit the number of countries a small business may sell its product. The potential consequences for limiting the market for certain technologies may include fewer firms pursuing development efforts for that technology and fewer investors providing funding for companies pursuing that particular technology. These are just a couple of examples of potential unintended consequences when an emerging technology is limited by law or regulation.[106] (This is taken up again in the discussion under Barrier 11 regarding export controls).

Best Practices

Recent Executive Action: On February 20, 2014, President Obama announced three new initiatives to encourage innovation and strengthening the patent system. Those initiatives, which will be implemented by the U.S. Patent and Trademark Office, are crowdsourcing prior art, having more robust technical training and expertise at the USPTO, and providing patent pro bono and pro se assistance.[107]

AM partnership meetings: For the last two years, the USPTO has hosted AM partnership meetings at its headquarters in Alexandria, Virginia. The goal of the program is for industry leaders, USPTO personnel and other stakeholders to exchange ideas and experiences about the state of the AM industry. This type of engagement should lead improved patent reviews by USPTO. Similar USPTOindustry working groups can be established for other emerging technologies.

Big Ideas–Recommendations

Technology Commercialization Assistance Fund: An investment from this fund could be used to pay patent filing expenses. See discussion under of the fund under Barrier 5 regarding access to capital.

Expand pro bono services: Allow small companies to take advantage of the USPTO's pro bono services in connection with the defense of the company's IP. Small firms could use the service if another company was accusing the small firm of patent infringement or the small firm wanted to prosecute an infringement claim against another company. Small companies are vulnerable to patent infringement and, in many cases, unable to defend or prosecute such claims due to the expense of patent litigation. The patent offers no benefit if the company is unable to enforce it.

Approve reasonable legislation dealing with patent trolls: The Innovation Act (HR 3309) was passed by the House of Representatives in December 2013.[108] The bill seeks to combat the issue of patent trolls, which are companies that don't create or sell any products but acquire patents for the purpose of initiating patent infringement lawsuits against firms using technologies possibly related to the patent. Several groups, including, had reservations about the bill due to the ramifications for small inventors.[109] The issue of patent trolls should be addressed by a new bill that does not negatively affect small inventors.

Barrier 11: Small Companies Continue to Face Challenges Exporting Their Products and Services

1 Discovery/ Identified Market Need	2 Technology Demonstration	3 Product Development	4 Commercialization/ Market Entry	5 Early Stage Growth	6 Economic Development Impact Growth
				Exporting	

What we heard: Small companies are becoming increasingly interested in and, in fact, participating more in exporting.[110] The most often heard issue was that it was still difficult for small innovative companies to find clear direction on what steps to take to begin exporting. Many small businesses we spoke with are unaware of the various government export assistance programs. Those businesses aware of the programs are often confused about which of the six agencies involved in exporting can provide the specific assistance needed for their firm (Department of Commerce, Export-Import Bank, Overseas Private Investment Corporation, Small Business Administration, U.S. Trade Development Agency, and the U.S. Trade Representative). This particular problem was recognized by President Obama when he sought authority to consolidate export activities of the different agencies.[111]

A related issue is that small businesses often do not have the resources (e.g., expertise and systems) to export because of its complex nature. Many small businesses must hire consultants to help with their exporting activities. Difficulty accessing specialized financial products to facilitate international sales was also noted. The Export–Import Bank does have small business products that provide for faster administrative processing and financial support for firms exporting. Another barrier noted as constraining exports was transportation costs. Small firms generally are not shipping in bulk and therefore their transportation cost structure is not as favorable as it is for large corporate exporters. The cumulative effect of these various expenses is that some small firms determine that exporting is too complex, expensive and not profitable so they decide not to do it.

Export controls can have significant impacts on small businesses and on technology industries. The potential impact can be seen by looking at export controls and AM. There is a possibility that the government ma place export controls on AM equipment, materials, and processes related to the production of metal parts/products. Export controls may have an adverse impact on the metal AM sector for two reasons: firms will be limited to exporting only to signatories to the Wasennaar Agreement[112] (i.e., limited export markets) and U.S. firms will be required to obtain a license to export to Wasennaar countries. Current licensing times are range from 30 to 120 days or more, which is longer than in other member countries.

Export controls impede all companies in transacting business. Controls may be warranted in cases where the technology has potentially special or sensitive applications and limited foreign availability. This is not the case for AM. In fact, there are already experts and manufacturers of AM industrial systems that make metal products in a number of Wasennaar countries (e.g., Germany, France) and in countries not party to the treaty (e.g., China).

The effect on small businesses is often disproportionate when export controls are implemented. Small firms cannot afford to have compliance systems in place to manage their licensing, tracking, and reporting responsibilities like large companies. Small innovative companies are further burdened by the slow pace of export licensing determinations. Many firms simply cannot afford the carry costs of a delayed sale while they wait to be informed whether or not they will receive a license to export the product. Further, small companies are at risk of the customer purchasing from another country's manufacturer whose licensing process is faster or from another country without any export restrictions. They also run the risk of their product becoming technically obsolescent during the approval period.

There have been occasions where export controls have had the unintended consequence of negatively affecting entire U.S. industries such as satellites and five axis machine tools.[113] The U.S. is a leader in the growing AM technology industry. Hopefully that leadership will not be derailed by export controls.

Best Practices

Export Control Initiative: Launched in 2009, the Export Control Initiative attempts to simplify and make compliance requirements less burdensome under export laws and regulations by reviewing, reclassifying and combining into one list items subject to export controls.[114]

Export.gov: Export.gov is the federal government's export promotion and finance portal. The portal is targeted to small and medium-sized U.S. companies to provide them with information on exporting and export services.

The portal combines information from many U.S. agencies and some private sector companies and is meant to be a starting place for companies interested in starting or expanding their exporting activities.

State Trade and Export Promotion (STEP) Grant Program: The STEP program is a three-year pilot program that makes matching grants to states to assist small businesses begin and succeed in exporting. Services under the program are provided by state government organizations and administered at the national level by the SBA. Some of the services offered through STEP include supporting small business participating in foreign trade shows and foreign market sales trips, website translation, assistance in international marketing efforts, subscription to services provided by the Department of Commerce, export trade show exhibits, training, and other efforts aligned with program goals. There is broad support from the small business community to make STEP a permanent program.[115]

National Export Initiative/Next (NEI/NEXT): In 2010, the President initiated a government-wide effort to assist U.S. businesses increase exports and expand into new markets called the National Export Initiative (NEI). NEI has helped increase U.S. exports for four consecutive years reaching an all-time high of $2.3 trillion in 2013 and supporting 11.3 million jobs.[116] Building on the success of NEI, the Administration announced the launch of NEI/NEXT in May 2014. NEI/NEXT is a data-based, customer service-driven initiative designed to further support U.S. companies exporting by:[117]

- providing export assistance to connect with overseas customers,
- facilitating trade by streamlining government processes and reporting requirements to make export shipments easier and less expensive,
- expanding access to export financing,
- expanding state and local partnerships to include exporting and attracting foreign investment as part of their economic development strategy, and
- working with foreign economies to improve their business environment and open markets so that American companies can more easily export abroad.

Big Ideas - Recommendations

Promotional campaign for federal trade agencies: The federal agencies with trade responsibilities and *Export.gov* provide much-needed export assistance to small firms via their respective websites, but based on our outreach, this information is not getting out to small businesses. The agencies should create and execute a promotional campaign to advise small businesses about their programs and capabilities. Higher utilization of these agency tools will help small innovative companies export.

Export counselors: The government should create a division of export counselors that are cross-trained with in depth knowledge about all government export programs that can assist small businesses. These counselors can serve as an entry point for small businesses ready to begin exporting that have submitted a profile detailing company information and exporting goals. *Export.Gov* is a great portal for entrepreneurs to learn about exporting. However, given its complex nature, many entrepreneurs would likely prefer being able to discuss their plans with a live person who can provide direct assistance.

Cross training allows the counselors to assess the needs of the company and recommend the proper program(s) at the appropriate agency or agencies. This centralized entry point for exporting would allow the agencies to decrease the level of their individual outreach activities. In addition, those programs that are being utilized most can be expanded or improved while those that are not can be revised or eliminated.

The program should use customer relationship management software (CRMS) to track companies that apply and use government programs to export. The CRMS will allow any counselor or any program manager at an export agency to see a history of federal export assistance a company has received and review the company's progress. This will allow government personnel to direct businesses to get the proper resources based on the company's stage of business development and export experience. The CRMS will also allow the government to capture data which can be used to improve its export programs and serve small businesses better. The CRMS must be carefully designed to allay the privacy concerns of the participating small businesses.

CONCLUSION

The Office of Advocacy's Innovation Initiative was charged with identifying the challenges and barriers hindering the growth and development of small innovative companies, and offering actionable solutions that may be implemented by or in conjunction with the government to surmount such barriers. The 11 barriers cited were never intended to be an exhaustive list of the barriers we heard about from the entrepreneurs and organizations interviewed. They are, however, the most often cited barriers, and collectively, they illustrate the complexity of the challenges to small innovative company growth and their dynamic, multidisciplinary nature. This broad examination of 11 barriers to innovative company growth has highlighted dozens of institutions and programs that are worth supporting and replicating in the goal of strengthening America's innovation ecosystem. Two additional general recommendations cut across most all of these areas toward the goal of facilitating small innovative company growth.

Pilot programs: For those agencies involved in the innovation process, a small portion of their budgets should be reserved to allow agency heads to implement initiatives or pilot programs to facilitate innovation or innovative company growth. Reserving funds for pilot programs gives agency heads latitude to experiment and try new things and will help eliminate inaction due to fear of failure. Innovations don't always work, and these pilots won't always be successful. However, fostering an environment that encourages new ideas and risk taking will benefit federal agencies seeking to assist innovative companies.

Coordination of innovation initiatives: Fostering an environment where small innovative companies can grow is a complex undertaking. It involves multiple disciplines including education, R&D, workforce development and training, finance, and law. If we believe that innovative companies are important drivers of our economy, then we need to increase the coordination of government resources, programs and initiatives to ensure a robust innovation ecosystem exists in the United States. President Obama recognized the need for greater efficiency in 2011 with his Government Reform for Competitiveness and Innovation Initiative.[118]

Other commentators have echoed the need for increased coordination of government innovation efforts. The Center for American Progress put forth a proposal in 2012 to streamline and integrate various government programs and infrastructure to focus on U.S. competitiveness and innovation.[119] Similarly, an expert panel looking at challenges to commercializing federally funded R&D recommended the creation of a new Office of Innovation and Federal Technology Partnerships. This office would coordinate, implement, and monitor government efforts to commercialize federal R&D.[120] The panel stated that this office "must be placed at the highest level [within government] to affect its multi-agency authority and oversight." Their suggestion was to place it within the Office of Management and Budget or create a new cabinet-level position, the U.S. Secretary for Innovation.

The government should also perform an extensive review of its own activities affecting innovation. This review will reveal opportunities for increased efficiency (e.g., eliminating duplicative or outdated programs). Next, the programs should be tracked for their respective impact on innovation and the development and growth of innovative companies. Tracking these programs' impact will create a large amount of data which should be analyzed for insights into the U.S. innovation ecosystem. With this knowledge, the White House, agency heads, and Congress will be able to better coordinate government innovation efforts, improve government programs, and allocate resources more effectively.

This report represents the findings of outreach to hundreds of individuals, businesses, and institutions. We hope it will be widely read within government and in the broader innovation ecosystem. We welcome all input at *advocacy@sba.gov*. The Office of Advocacy will continue to listen to America's small innovators and voice their concerns within government.

Other Reading

Robert D. Atkinson, Understanding the U.S. National Innovation System, The Information Technology and Innovation Foundation, June 2014.

Bourell David L., Rosen David W., and Leu Ming C. (eds.), Roadmap for Additive Manufacturing: Identifying the Future of Freeform Processing, The University of Texas at Austin, Laboratory for Freeform Fabrication, Advanced Manufacturing Center, 2009.

Thomas Campbell, Christopher Williams, Olga Ivanova, Banning Garrett, Could 3D Printing Change the World? Technologies, Potential, and Implications of Additive Manufacturing, The Atlantic Council, October 2011.

Daniel Cohen, Matthew Sargeant,and Ken Somers, 3D printing takes shape, McKinsey Quarterly, January 2014.

Scott Crump, Direct digital manufacturing part three: How to identify opportunities, Stratasys, accessed September 11, 2014. http://www.stratasys.com/resources/~/media/09545949F30740EBA3A3D4746D093224.pdf

Helen Delaney, Choosing Standards Based on Merit: Liberalizing Regulation, Trade and Development, ASTM International, 2010. http://www.astm.org/ABOUT/images/Choosing_Standards.pdf

John R. Dearie and Courtney C. Geduldig, Where the Jobs Are: Entrepreneurship and the Soul of the American Economy, (Hoboken, N.J.: John Wiley & Sons, Inc.) 2013.

Nora Freeman Engstrom, 3-D Printing and Product Liability: Identifying the Obstacles, 162 University of Pennsylvania Law Review Online 35, 2013.

Ewing Marion Kauffman Foundation, State of Entrepreneurship Address: Towards America's New Entrepreneurial Growth Agenda, Ewing Marion Kauffman Foundation, February 12, 2014.

John F. Hornick, 3D printing and the future (or demise) of intellectual property, 3D Printing and Additive Manufacturing. March 2014, 1(1): 34-43.

Barry Jaruzelski, John Loehr, and Richard Holman, The Global Innovation 1000: Navigating the digital future, Strategy + Business Magazine, October 2013.

Kevin Jurrens, Measurement Science for Additive Manufacturing, National Institute of Standards and Technology–U.S. Department of Commerce, March 2013. http://www.nist.gov/el/msid/infotest/upload/Jurrens-NIST-AM_Projects_March2013.pdf

Lagorio-Chafkin, Christine, "Clay Christensen: The wrong kind of innovation," Inc. Magazine, October 2, 2013. http://www.inc.com /christine-lagorio/clayton-christensen-capitalist-dilemma.html

McKinsey Global Institute and Mckinsey Operations Practice, Manufacturing the Future: The Next Era of Global Growth and Innovation, McKinsey & Company, 2013.

McKinsey Global Institute, Game changers: Five opportunities for U.S. growth and renewal, McKinsey & Company, July 2013.

National Economic Council, Council of Economic Advisers and Office of Science and Technology Policy, A Strategy for American Innovation: Securing Our Economic Growth and Prosperity, February 2011.

National Science Foundation, National Center for Science and Engineering Statistics, Science and Engineering Indicators 2014, Arlington, VA (NSB 14-01), February 2014.

Ben Olinsky and Sarah Ayers, Training for Success: A Policy to Expand Apprenticeships in the United States, Center for American Progress, December 2013.

The OECD and the World Bank, The Innovation Policy Platform website, https://www.innovationpolicyplatform.org/

Ivan Pope, The coming 3D ecosystem: Part 4, 3D Printing Industry, July 2014, accessed September 12, 2014. http://3dprintingindustry.com/2014/07/07/coming-ecosystem-3d-printing-part-4/

PricewaterhouseCoopers, Innovation: Government's Many Roles in Fostering Innovation, PricewaterhouseCoopers, 2010.

PricewaterhouseCoopers, Rethinking Innovation in Industrial Manufacturing: Are You Up for the Challenge? PricewaterhouseCoopers Global Innovation Survey 2013: Industrial Manufacturing Perspectives, 2013.

PricewaterhouseCoopers, Breakthrough Innovation and Growth, PricewaterhouseCoopers Global Innovation Survey 2013, 2013.

Justin Scott, Nayanee Gupta, Christopher Weber, Sharon Newsome, Terry Wohlers and Tim Caffrey, Additive Manufacturing: Status and Opportunities, IDA Science and Technology Policy Institute, March 2012.

Technology CEO Council, High Impact: How IT is Empowering the Next Generation of Entrepreneurs, Technology CEO Council, 2012.

U.S. International Trade Commission, Small and Medium Sized Enterprises: U.S. and EU Export Activities, and Barriers and Opportunities Experienced by U.S. Firms, USITC Publication 4169, July 2010.

Michael Weinberg, What's the deal with copyright and 3D printing? Public Knowledge blog, white paper, January 2013.

Charles W. Wessner, ed., Committee on Capitalizing on Science, Technology, and Innovation: National Research Council, An Assessment of the Small Business Innovation Research Program, (Washington D.C.: The National Academies Press) 2008.

Garrett White and Daniel Lynskey, Economic Analysis of Additive Manufacturing for Final Products: An Industrial Approach, University of Pittsburgh-Swanson School of Engineering, April 13, 2013.

About the Author: Mark K. Harrison, Office of Advocacy Entrepreneur in Residence

Mark K. Harrison joined the Office of Advocacy in summer 2013 as its entrepreneur in residence. Harrison's background combines experience in government, the private sector, and law. Prior to joining Advocacy, he was president of the Business Consortium Fund, a not-for-profit corporation affiliated with the National Minority Supplier Development Council. The fund provides financing, business advisory services, and educational programs to minority entrepreneurs. In that capacity, he developed and implemented two new financing programs, an advisory service, and an educational program.

Harrison served as deputy director of the Investment Financing Group (IFG) for the Maryland Department of Business and Economic Development in Baltimore; there he managed programs making venture capital investments in early and seed stage technology companies. He served as a Maryland assistant attorney general and legal counsel to the IFG from 1994 to 1999 and worked with the U.S. Securities and Exchange Commission from 1990 to 1994.

Harrison earned his law degree from the University of Maryland School of Law in Baltimore and a bachelor of science in business and marketing management from Cornell University.

End Notes

[1] National Research Council, Rising to the Challenge: U.S. Innovation Policy for the Global Economy (Washington D.C.: The National Academies Press 2012) 201-319.

[2] IEEE is a professional association of persons working in technology fields whose mission is to foster technological innovation and excellence for the benefit of humanity. The IEEE survey had 240 respondents; 90% owned their own business, 99% had less than 99 employees, and 20% were owned by a minority, woman, or veteran.

[3] The White House, A Strategy for American Innovation–Securing Our Economic Growth and Prosperity, February 2011.

[4] PriceWaterhouseCoopers, Breakthrough Innovation and Growth, September 2013.

[5] OECD, 2005, The Measurement of Scientific and Technological Activities: Guidelines for Collecting and Interpreting Innovation Data: Oslo Manual, Third Edition, prepared by the Working Party of National Experts on Scientific and Technology Indicators, OECD, Paris, para. 146, accessed December 22, 2014.

[6] Robert Litan, Andrew W. Wyckoff, and Kaye Husbands Fealing, eds. Capturing Change in Science, Technology and Innovation: Improving Indicators to Inform Policy, (Washington, D.C., The National Academies Press) 2013.

[7] Jake Neilson, Four types of innovation and the strategic choices each one represents, Innovation Excellence blog, January 12, 2014.

[8] Clayton Christensen, We are living the capitalist's dilemma, CNN Website, January 21, 2013. Accessed August 20, 2014.

[9] U.S. Small Business Administration, Office of Advocacy, Frequently Asked Questions, March 2014. Office of Advocacy estimate.

[10] Hathaway, Ian, Tech Starts–High Technology Business Formation and Job Creation in the United States, The Kauffman Foundation, August 2013.

[11] The Executive Office of the President, Making in America: U.S. Manufacturing, Entrepreneurship and Innovation, June 2014.

[12] National Science Foundation Business R&D and Innovation Survey (BRDIS) 2008.

[13] The Executive Office of the President, Making in America.

[14] McKinsey Global Institute, Disruptive Technologies: Advances That Will Transform Life, Business, and the Global Economy, May 2013. Others have estimated the additive

manufacturing market to reach between $21 billion and $200 billion. Wohlers Associates, The Wohlers Report 2014.

[15] Wohlers Associates, The Wohlers Report 2014. This report is the source of most of the background and statistics on the additive manufacturing process and industry in this section.

[16] As defined by the ASTM International Committee F-42 on Additive Manufacturing, a standards development organization.

[17] According to the Wohlers Report 2014, 29% of survey respondents are using additive manufacturing for finished or functional parts; 37% for prototyping/modeling ; 5.6% for tooling components; and 6% for education/research..

[18] This is a bit misleading in that Stratasys, a U.S. based company which had 54.7% of global unit sales of industrial systems, merged with an Israeli company in December 2012 with the new entity choosing to register as an Israeli company. Stratasys maintains significant operations in the U.S. Prior to 2013, the U.S. produced almost 61% of industrial systems. Wohlers Associates, The Wohlers Report 2014.

[19] Stratasys acquired two service providers in 2014 (Harvest Technologies and Solid Concepts) and 3D Systems, a publicly traded U.S.-based 3D printing system manufacturer, acquired Medical Modeling as well as Robtec which is based in Brazil. The Wohlers Report 2014.

[20] Estimated growth rate of sales of personal 3D printers was 171% over the last 3 years. The Wohlers Report 2014.

[21] There are efforts being made to develop a personal 3D printers that use metals as build material. Michael Molitch-Hou, A sub-$4,000 metal 3D printer? 3D Printing Industry, September 23, 2014. Michael Molitch-Hou, Michigan Tech releases open-source 3D metal printer for less than $2,000? 3D Printing Industry, December 9, 2013.

[22] White House Maker Faire website, June 2014.

[23] NASA, NASA tests limits of 3-D printing with powerful rocket engine check, NASA blog, August 2013.

[24] Helmuth Ludwig and Eric Spiegel, America's real manufacturing advantage, *strategy + business*, January 20, 2014, accessed August 20, 2014.

[25] John Barber and Jim Tully, Emerging Market Opportunity: The Internet of Things, Gartner, Inc., June 24, 2014 (webinar presentation).

[26] Researchers at Microsoft have developed a technology, called InfraStructs, that allows information to be embedded into 3D printed objects during the production process that can be read by external scanners. This is similar to radio frequency identification codes (RFID) except that no security cards or keys are required because the information is embedded in the object. Rakesh Sharma, Microsoft researchers introduce technology for an internet of things in 3D printing, Forbes Magazine, September 12, 2013, accessed August 20, 2014.

[27] Greg Gorbach and Andy Chatha, Information-Driven Manufacturing, ARC Strategies, February 2013. Ludwig and Spiegel, America's Real Manufacturing Advantage.

[28] Joseph Bradley, et. al., The Internet of Everything: Top 10 Insights from Cisco's IoE Value at Stake Analysis for the Public Sector, Cisco Systems, 2013, accessed August 20, 2014.

[29] Ludwig and Spiegel, America's Real Manufacturing Advantage, page 48.

[30] Adopted from Deborah Jackson, What is an Innovation Ecosystem? (working paper 2011) Arlington, VA, National Science Foundation, and Byron Clayton, "Scaling, Accelerating and Sustaining the Innovation Pipeline," (presentation and interview at NorTech headquarters, Cleveland, Ohio, February 19, 2014).

[31] Ralph Landau and Nathan Rosenberg, The Positive Sum Strategy; Harnessing Echnology for Economic Growth, (Washington, D.C., The National Academies Press) 289-294, 1986.

[32] Landau and Rosenberg, The Positive Sum Strategy.

33 Jackson, What is an Innovation Ecosystem? Elgar, Capturing Change in Science, Technology and Innovation.

34 Christopher L. Weber, et. al, The Role of the National Science Foundation in the Origin and Evolution of Additive Manufacturing in the United States, IDA Science & Technology Policy Institute, November 2013.

35 The barriers were those identified in the AM industry that we believe are applicable to many small innovation firms. Thus, they are set forth as general statements not specifically barriers to growth in AM.

36 U.S. Small Business Administration, Office of Advocacy, Student debt among young entrepreneurs, fact sheet, November 2014. Federal Reserve Board, Survey of Consumer Finances, 2013.

37 Federal Reserve Bank of New York, Student Loan Debt by Age Group, March 29, 2013.

38 Office of Advocacy, Frequently Asked Questions about Small Business, March 2014.

39 Merra Lousi, Student debt puts young entrepreneurs on hold, Businessweek, June 20, 2013; Phyllis Korkki, The ripple effects of rising student debt, New York Times, May 23, 2014,

40 The White House blog, Income based repayment: everything you need to know.

41 Michael B. Farrell, Rhode Island may defer loans for young entrepreneurs, Boston Globe, August 27, 2013, accessed August 20, 2014.

42 Venture for America website; VFA Opportunity Awards Guidelines.

43 Total U.S. R&D expenditure was $424.4 billion in 2011 with business and the federal government funding $267.2 billion and 125.7 billion respectively. NSF Science and Engineering Indicators 2014, Table 4-1.

44 Businesses do little basic research because the results are often far removed from product development. Businesses spend their time and money on creating products that can improve their bottom line. Spending on applied research and development increased from 0.6% of GDP in 1953 to 2% of GDP in 2011.

45 NSF Science and Engineering Indicators 2014, Table 4-1.

46 Batelle, 2014 Global R&D Funding Forecast, R&D Magazine, December 2013.

47 Batelle, 2014 Global R&D Funding Forecast.

48 Batelle, 2014 Global R&D Funding Forecast. Batelle projects the U.S. will spend $465 billion in 2014 which will result in direct employment of 2.7 million U.S. residents in the public and private sectors and supporting an additional 6 million U.S. jobs .

49 White House Office of Science and Technology Policy, The 2015 Budget: Science, Technology, and Innovation for Opportunity and Growth, March 2014.

50 White House Office of Science and Technology Policy, the 2015 Budget. Basic research would decrease from $32.4 billion to $32 billion, applied research would increase from $32 billion to 32.6 billion, and development would increase from $66.5 billion to $68 billion.

51 The Global Innovation Initiative website.

52 H.R. 4438, the American Research and Competitiveness Act which was approved by the House of Representatives in May 2014, would give companies a 20% credit on a portion of their research expenses that exceeds their average ongoing research spending over the previous few years and makes the credit permanent.

53 The Expiring Provisions Improvement Reform and Efficiency (EXPIRE) Act (S. 2260) was sent to the Senate for consideration in April 2014. The bill extends for two years the 20% traditional research tax credit and the 14% alternative simplified credit. It also allows startup businesses to claim unused credits against their payroll tax after applying the credit to income tax liability. The benefit is capped at $250,000 per year and available only to

companies less than five years old with less than $5 million in gross receipts. U.S. Senate Finance Committee, Final Summary of the Expire Act as reported.

[54] The U.S. R&D tax credit rate is a maximum of 20% with a top corporate tax rate of 35% compared to other countries: Australia – tax credit rate is 45% for companies with revenue under AU.S. $20 million and corporate tax rate of 30%, Ireland – tax credit rate is 35% and corporate tax rate 12.5%. China, Japan, Singapore and the UK have corporate tax rates of 25%, 28%, 17%, and 23% respectively as well as combinations of other tax incentives such as super-deductions on R&D expenditures, tax credits, and patent box rates on certain IP products developed in country. Laughlin Cutler, et. al, Global R&D incentives compared, Journal of Accountancy, June 2013.

[55] National Science Foundation, I-Corps program website.

[56] The I-Corps curriculum is built on a special, accelerated version of Stanford University's Lean LaunchPad course with additional elements designed just for I-Corps grantees. The Lean LaunchPad curriculum was developed and is currently taught by serial entrepreneur Steve Blank.

[57] The NSF website expands on these: the five I-Corps nodes and I-Corps sites.

[58] The NSF and NIH Biomedical Pilot Program website.

[59] BusinessUSA.Gov website; Federal Laboratory for Technology Transfer website; SBIR and STTR program websites.

[60] Deloitte and the Manufacturing Institute, Boiling Point? The Skills Gap in U.S. Manufacturing, September 2011.

[61] NSF Engineering Research Centers website.

[62] Pursuant to the Presidential Memorandum on Job-Driven Training for Workers, Vice President Biden led a review of the U.S. workforce and job training programs and developed recommendations to make system more job-driven. For information, see: The White House, Ready to Work: Job-Driven Training and the American Opportunity, July 2014.

[63] See Department of Labor, Registered Apprenticeships–Frequently Asked Questions, accessed August 15, 2014.

[64] See Department of Labor, Registered Apprenticeship College Consortium–Frequently Asked Questions, accessed August 15, 2014.

[65] See the Department of Labor WOIA Resource Page and Department of Labor WOIA Fact Sheet.

[66] Senator Coons announces bill to boost manufacturing education at universities, Senator Coons' blog, accessed August 15, 2014. Robert Atkinson, Manufacturing universities: A catalytic step toward revitalizing American manufacturing, Industry Week, accessed August 15, 2014. The bill did not pass (see Govtrack.us website).

[67] One such effort is the Minnesota Advanced Manufacturing Workforce Pilot, led by South Central Community College. This group of 24 community colleges and employers is developing scalable apprenticeship models in mechatronics, as well as computer integrated machining and welding (accessed August 20, 2014).

[68] The National Institute of Standards and Technology's Hollings Manufacturing Extension Partnership (MEP) is a national network of centers located in all 50 states and Puerto Rico that works with small and mid-sized U.S. manufacturers to help them increase their efficiency by providing services focused on productivity improvement, product and workforce development, business practices and technology transfer.

[69] There are 60 MEP centers across the country with 5-7 centers with AM systems for prototyping.

[70] Startups may get financing from SBIRs/STTRs or other federally funded government research grants. Some states also have programs that provide funding for startups. In the private sector, startups may get funding from angel investors, VC funds focused on startups, or their own savings (self-financing).

[71] The micro VC glut, CB Insights, May 6, 2014.

[72] Micro VC rising: Analyzing trends and the top investors in the micro VC ecosystem, CB Insights, June 19, 2014.

[73] VCs warm up to 3D printing, CB Insights, April 22, 2014.

[74] SBICs are privately owned and managed investment funds, licensed and regulated by SBA, that use their own capital plus potential additional funding from the SBA in the form of SBA guaranteed debentures to make equity and debt investments in small businesses. The SBA does not invest directly into small business through the SBIC Program, but provides funding to qualified investment management firms with expertise in certain sectors or industries.

[75] The Maryland Venture Fund website; the Innovate New York Fund website; the Michigan Pre-Seed Fund Loan Program.

[76] NSF, Science and Engineering Indicators 2014, U.S. Bureau of Economic Analysis, U.S. International Services–CrossBorder Trade in 2012 and Services Supplied Through Affiliates in 2011, October 2013. Index Mundi, United States-Royalties and Licensing Fees.

[77] UK Intellectual Property Office, Banking on IP? The role of Intellectual Property and Intangible Assets in Facilitating Business Finance, 2013.

[78] Kenan Jarboe, Intangible assets innovative financing for innovation, Issues in Science and Technology, November 2013.

[79] Jeffery Sohl, The Angel Investment Market in 2013: A Return to Seed Investing, Center for Venture Research, April 2014.

[80] An accredited investor is currently defined as an individual with $1 million in net assets excluding their primary home, or income of $200,000, if single and $300,000 if married.

[81] SEC Release No. 33-9415 Eliminating the Prohibition Against General Solicitation and General Advertising in Rule 506 and Rule 144A Offerings, Economic Analysis.

[82] Sohl, Angel Investment Market in 2013. In 2013, minority-owned companies represented 7% of firms pitching to angel investors with 13% receiving funding. Women-owned companies represented 23% of firms pitching to angel investors with 19% receiving investment. The investment yield rate has been high the last two years, averaging 21.5%, but traditionally it has been around 15%.

[83] SBIR Phase III is the commercialization phase, in which a project moves from prototype to production through agency contracts for the specific item. The SBIR website describes it as follows: "The objective of Phase III, where appropriate, is for the small business to pursue commercialization objectives resulting from the Phase I/II R/R&D activities. The SBIR program does not fund Phase III. Some Federal agencies, Phase III may involve follow-on non-SBIR funded R&D or production contracts for products, processes or services intended for use by the U.S. Government."

[84] Michael B. Staebler, Description of the small business investment company program participation by funds using debentures, November 1, 2013, accessed August 15, 2014.

[85] See DARPA Small Business Planning Tool website; DOE Commercialization Assistance Program website; and NIH Niche Assessment Program website.

[86] Phillip Shipira, Stuart Rosenfeld, An Overview of Technology Diffusion Policies and Programs to Enhance the Technological Absorptive Capabilities of Small and Medium

Enterprises, background paper prepared for the OECD Directorate for Science, Technology and Industry, August 1996.

[87] Daniel L. Cohen, Fostering mainstream adoption of industrial 3D printing: Understanding the benefits and promoting organizational readiness, 3D Printing, vol. 1, no. 2, 2014.

[88] The ASTM International Technical Committee F42 on Additive Manufacturing Fact Sheet.

[89] NIST's Manufacturing Extension Partnership website.

[90] MEP's Technology Scouting Program website.

[91] Wohlers Associates, The Wohlers Report 2014, page 111.

[92] For example 3D Hubs based in Holland and Makexyz based in New York.

[93] OECD, Innovation Procurement Schemes, The Innovation Policy Platform, accessed August 20, 2014.

[94] In the effort to meet the annual small business goal of 23% of federal procurement, small firms are awarded some $80 billion in federal procurement awarded annually.

[95] OECD, Innovation Procurement Schemes citing Tsipouri, L., et al., (2010): "Risk Management in the Procurement of Innovation," Report of an Expert Group for the EU Commission, Brussels.

[96] The RFP-EZ website.

[97] See the Challenge.gov website.

[98] L. Elaine Halchin, Other Transaction (OT) Authority, Congressional Research Service, August 2011.

[99] GAO, Report to Congressional Committees, Homeland Security–Further Action Needed to Improve Management of Special Acquisition Authority, May 2012. Information on OTs is difficult to track due to a lack of specific reporting obligations but it appears that some agencies have been using it with traditional contractors (i.e., large companies). While there is no quantitative evidence that OT increases the use of non-traditional contractors, most contracting managers believe it is an effective tool. Halchin, Other Transaction Authority.

[100] Certain other agencies may use OT under certain conditions and if approved by the director of the Office of Management and Budget. Halchin, Other Transaction Authority.

[101] *3D Systems vs. Formlabs* in 2012, and *Stratasys vs. Microboards Technology (dba Afinia)* in 2013. Each case involves a publicly traded manufacturer of AM systems and 3D printers suing a smaller company producing or distributing a different brand of 3D printer. The Stratasys case is of note because it is claiming infringement on patents regarding improvements to fused deposition modeling technology which is the technology used by almost all 3D printer manufacturers. Should Stratasys prevail, there are potential huge consequences for the 3D printer industry. See, Michael Weinberg, Afinia responds to Stratasys: Your patents are invalid and your threats are anticompetitive, Public Knowledge Blog, January 3, 2014.

[102] For good discussions on this topic, see Melba Kurman, Carrots, not sticks: rethinking enforcement of intellectual property rights for 3D-printed manufacturing, 3D Printing vol.1 no.1, 2014,; Michael Weinberg, It will be awesome if they don't screw it up: 3D printing, intellectual property, and the fight over the next great disruptive technology, Public Knowledge Blog white paper, November 2010.

[103] Nora Freeman Engstrom, 3-D printing and product liability: identifying the obstacles, University of Pennsylvania Law Review Online.

[104] Gartner.com predicts 3D printing will result in the loss of at least $100 billion per year in intellectual property globally by 2018.. Given the development of the technology at this point, the most vulnerable markets would one where the products are small and made of plastic (e.g., toys). The counterfeiting of metal products is unlikely due to the expense to

produce product (i.e., AM system, materials, etc.), the expertise required to build parts and the limited number of AM metal product manufacturers and the relatively small supply chain for such products.

[105] Cameron Naramore, Could your 3D printer someday spy on you? 3Dprinter.net, October 2012.

[106] Michael Weinberg, Undetectable firearms law passes without 3D printing, Public Knowledge blog. Recounts how restrictions were almost placed on AM technology in connection with the renewal of the Undetectable Firearms Act.

[107] On June 4, 2013, President Obama announced five executive actions "to help bring about greater transparency to the patent system and level the playing field for innovators." For a full description of all executive actions see http://www.uspto.gov/patents /init_events /executive_actions.jsp#heading-6

[108] The Innovation Act (HR 3309) summary.

[109] See letter from Chief Counsel for Advocacy Winslow Sargeant to Sen. Mary Landrieu dated March 12, 2014.

[110] Small and medium-size enterprises (with fewer than 500 employees) represent almost 98% of the 302,360 firms that exported in 2011. Similarly SMEs made up 97% of U.S. manufacturers exporting representing 19% of the total value of manufacturing exports. Department of Commerce, International Trade Administration, U.S. Exporting Companies 2011.

[111] The White House, Government Reorganization Fact Sheet, January 13, 2012.

[112] "The Wassenaar Arrangement has been established in order to contribute to regional and international security and stability, by promoting transparency and greater responsibility in transfers of conventional arms and dual-use goods and technologies, thus preventing destabilising accumulations. Participating States seek, through their national policies, to ensure that transfers of these items do not contribute to the development or enhancement of military capabilities which undermine these goals, and are not diverted to support such capabilities. The decision to transfer or deny transfer of any item is the sole responsibility of each Participating State." There are 41 other countries besides the U.S. party to the arrangement.

[113] National Air and Space Intelligence Center, Briefing of the Working Group on the Health of the U.S. Space Industrial Base and the Impact of Export Controls, Center for Strategic and International Studies, February 2008 (satellites); Richard A. McCormack, U.S. precision machine tool industry is no longer a global competitive force, Manufacturing and Technology News, March 5, 2010.

[114] Export Control Reform Initiative website. Currently, the export of military items that are listed on the U.S. Munitions List (USML) is the responsibility of the State Department's Directorate of Defense Trade Controls pursuant to the International Traffic in Arms Regulations (ITAR). The export of certain commercial or duel use items that are listed on the Commercial Control List (CCL) is the responsibility of the Department of Commerce's Bureau of Industry and Security pursuant to the Export Administration Regulations. The Export Administration Regulations are, among other things, combining the USML and CCL into a single list of items subject to export control.

[115] State Trade and Export Promotion (STEP) Grant Program website, accessed August 15, 2014.

[116] Department of Commerce International Trade Administration, Fact Sheet, National Export Initiative/NEXT, accessed August 15, 2014.

[117] U.S. Trade Promotion Coordinating Committee and Export Promotion Cabinet, National Export Initiative/NEXT–Strategic Framework, May 13, 2014, accessed August 15, 2014.

[118] The White House, Presidential Memorandum on Government Reform for Competitiveness and Innovation, March 11, 2011; an initiative to streamline executive branch agencies in order to increase trade, exports, and U.S. competitiveness.

[119] Jonathan Sallet and Sean Pool, Rewiring the Federal Government for Competitiveness: A New Cabinet Department for the 21st Century, Center for American Progress, January 2012, accessed August 20, 2014.

[120] White House Lab-to-Market Inter-Agency Summit: Recommendations from the National Expert Panel, May 20, 2013, accessed August 20, 2014. This summit convened by the White House Office of Science and Technology Policy and the National Institutes of Health, National Heart, Lung and Blood Institute consisted of 20 federal agency representatives and outside experts who examined the challenges to commercialization of federal R&D and recommended solutions.

In: Small Business Innovators
Editor: Mary Brown

ISBN: 978-1-63483-265-6

Chapter 2

SUPPLY CHAIN INNOVATION: STRENGTHENING AMERICA'S SMALL MANUFACTURERS*

Executive Office of the President and the U.S. Department of Commerce

INTRODUCTION

U.S. manufacturing is in the midst of a potential resurgence. Since February 2010, manufacturers have added 877,000 jobs – the sector's first sustained job growth since the 1990s.[1] America's 230,000 small manufacturing firms[2] have played a key role in this resurgence, adding the majority of new manufacturing jobs every year, while forming the backbone of U.S. supply chains.

Dense networks of these small manufacturers are vital to the process of taking a product from concept to market, and the exchange of manufacturing know-how across suppliers is essential for the diffusion of the new technologies and innovative processes that give U.S. manufacturing its cutting edge.

However, these networks are under stress. From 2000 to 2010, manufacturing output and investment stagnated as companies offshored

* This is an edited, reformatted and augmented version of a report issued by the Economics and Statistics Administration, March 2015.

production previously done domestically. The U.S. supplier base weakened, raising doubts about the future of manufacturing's contribution to American innovation.[3]

Small manufacturers then and now face steep barriers to innovation – from invention, to commercialization, to the diffusion of new technology. For example, small manufacturers contribute less than one-third of all manufacturing R&D, despite making up 98 percent of manufacturing firms and as a general rule, small firms are one-seventh as likely as large firms to conduct R&D.[4]

As a result, small manufacturers are often in the positon of having to adopt technologies invented by others. Small manufacturers also face unique barriers in accessing the capital and expertise to take on the risk of new technologies. In part because of the challenges they face in adopting new technologies and processes, small manufacturers are only 60 percent as productive as large manufacturers.[5]

Strengthening America's supply chains and the small manufacturers at their core is essential to the long-term competitiveness of U.S. manufacturers both large and small. Manufacturers spend on average 60 percent of the price of their final product on purchased inputs, so differences in the quality and nimbleness of their supply chains can make or break a manufacturer's ability to compete.[6]

Large manufacturers have a critical role to play in cultivating the capabilities of small firms in their supply chains and spurring cross-pollination of expertise across firms. But many currently underinvest in their supply chains because of conflicting internal goals or the fear of strengthening small businesses that may also serve their competitors. By focusing on supplier quality and innovation rather than engaging in a race to the bottom on piece price, companies can send a strong market signal that stimulates innovation by small suppliers, thereby making the entire supply chain more competitive.

The public sector can augment and catalyze private-sector efforts such as those described above. Using its unique innovation assets, the government can help small manufacturers access state-of-the-art research, engineering expertise, and equipment that normally would be out of reach for any one small company. Government can also convene supply chain consortia to develop new technologies, and can highlight best practices in customer-supplier relationships.

For example, the Hollings Manufacturing Extension Partnership at the Department of Commerce is a network of 60 centers and 1,200 manufacturing experts across the country. By providing technical expertise to small manufacturers, the state-federal partnership strengthens the capabilities of individual suppliers and entire supply chains, spurs new linkages between suppliers, and provides small manufacturers with insight into technologies that can revolutionize their business. In addition, the Department of Energy's National Labs, leading national innovation assets, can help small manufacturers access cutting-edge manufacturing equipment, try out technologies and get answers to pressing research challenges through programs like those run by the Los Alamos and Sandia National Laboratories.

The United States has made considerable progress in bolstering American manufacturing since the Great Recession, but there is still more to be done to grow middle-class jobs and help businesses expand in this vital sector. The Administration's focus on reinvesting in the small manufacturers that comprise the core of America's supply chains is just one example of how — through sustained focus and investment across the public and private sectors — the United States can rebuild what was lost and lay a foundation to sustain and deepen the resurgence underway in U.S. manufacturing.

THE INCREASING IMPORTANCE OF SMALL MANUFACTURERS AND SUPPLY CHAINS

Dense networks of small- and medium-sized manufacturers increasingly power the engine of America's supply chains. For much of the 20th century, firms were likely to design and build their own products using internally produced parts and proprietary technologies developed by their own employees.[7] In recent decades, such vertical integration has become less common, with many firms focusing instead on a few core competencies and outsourcing other stages of production to suppliers, sometimes thousands of suppliers.

The automobile industry provides just one example. In the 1920s, '30s, and '40s, Ford Motor Company's River Rouge production complex was a mile-and-a-half wide and over a mile long, with 93 buildings and 15.8 million square feet of floor space. In addition to furnaces for making iron and steel, the complex "included a tire-making plant, stamping plant, engine casting plant, frame and assembly plant, transmission plant, radiator plant, tool and die plant, and, at one time, even a paper mill. A massive power plant produced enough electricity to light a city the size of nearby Detroit, and a soybean conversion plant turned soybeans into plastic auto parts." Ford also owned the natural resources needed to produce automobiles: 700,000 acres of forests, iron mines and limestone quarries, coal-rich land, and a rubber plantation.[8]

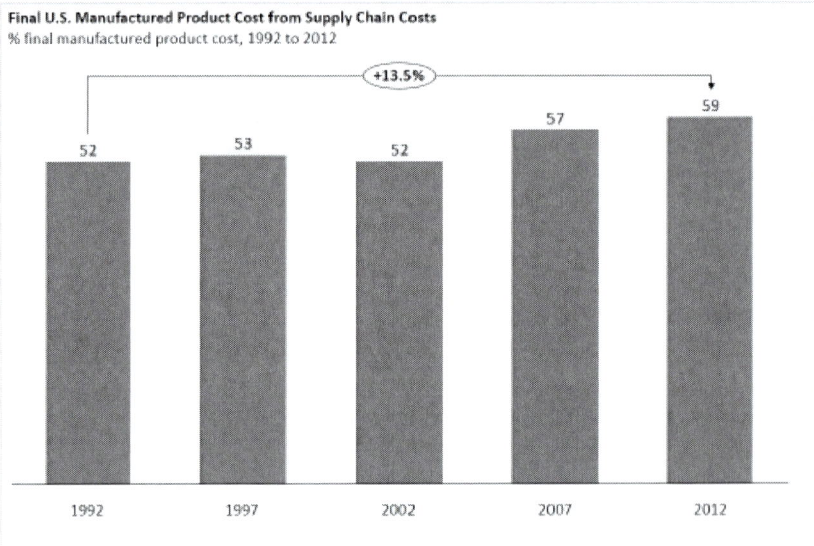

Source: Census Bureau's Economic Census (1992 to 2012).

Today, rather than making most of their inputs in company-owned factories or sourcing raw materials from company-owned plantations and mines, most automakers depend on outside firms to design and produce parts. Close to 60 percent of the purchase price of a new car today comes from value added by suppliers, including everything from cup-holders to semiconductors to steering wheels.[9] And the same is true for U.S. manufacturing overall.[10] Far more workers are now employed by suppliers than by the lead firm. For example, Chrysler's Toledo assembly complex employed 2,400 workers in 2011, but even more workers – 3,000 total – were employed in the 200 "Tier 1 suppliers" that shipped parts directly to the Toledo plant. These suppliers included a manufacturer in Dry Ridge, KY that provided the axles, another in Saranac, MI that made the door handles, and yet another factory in Northwood, OH that supplied the seats.[11] Some parts came from foreign countries, such as wiring harnesses from Mexico.

Chrysler's Tier 1 suppliers were, in turn, supplied by Tier 2 suppliers, who themselves were supplied by Tier 3 suppliers, pulling from industries as diverse as computer and electronics manufacturing, textiles, plastic and rubber products, and metal forming. When all the tiers of the supply chain are added up, an automaker may have thousands of suppliers.[12]

As firms shift from in-house production to more distributed production in networks of suppliers, the health of supply chains becomes more important to the competitiveness of both industries and nations. In addition, this shift towards networks of specialized suppliers has raised the importance of small manufacturers to overall growth and employment in manufacturing.

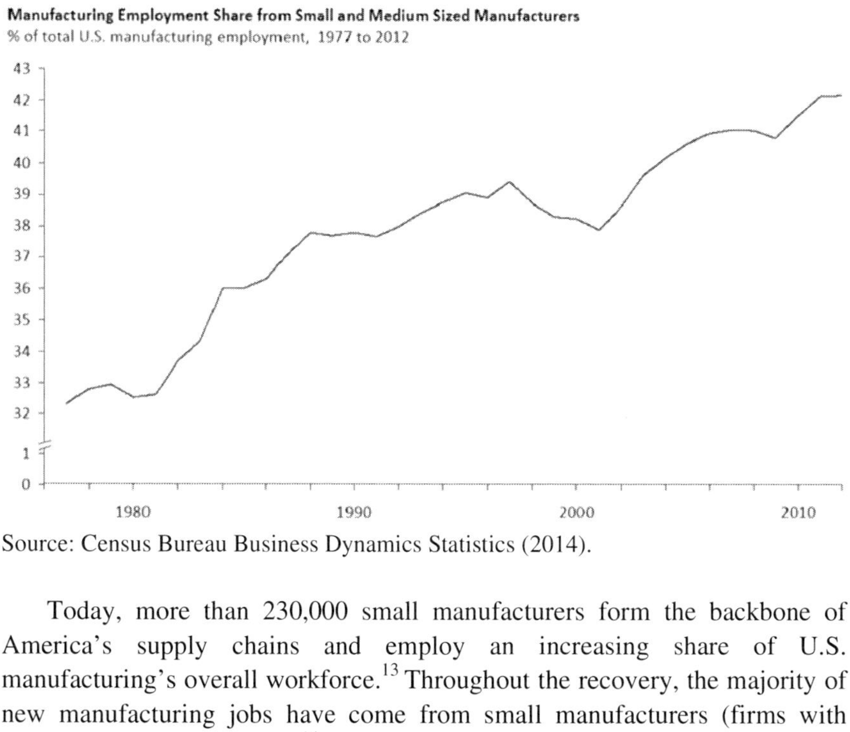

Source: Census Bureau Business Dynamics Statistics (2014).

Today, more than 230,000 small manufacturers form the backbone of America's supply chains and employ an increasing share of U.S. manufacturing's overall workforce.[13] Throughout the recovery, the majority of new manufacturing jobs have come from small manufacturers (firms with fewer than 500 employees).[14] In the 1980s, small manufacturers accounted for less than a third of all U.S. manufacturing employees. Today, small manufacturers employ 42 percent of all U.S. manufacturing workers, a steady increase of ten percentage points over three decades.[15]

Because small manufacturers play an increasing role in the overall manufacturing sector as members of tightly interdependent supply chains, their ability to keep up with and even lead advances in technology is critical to the competitiveness of U.S. manufacturing overall.

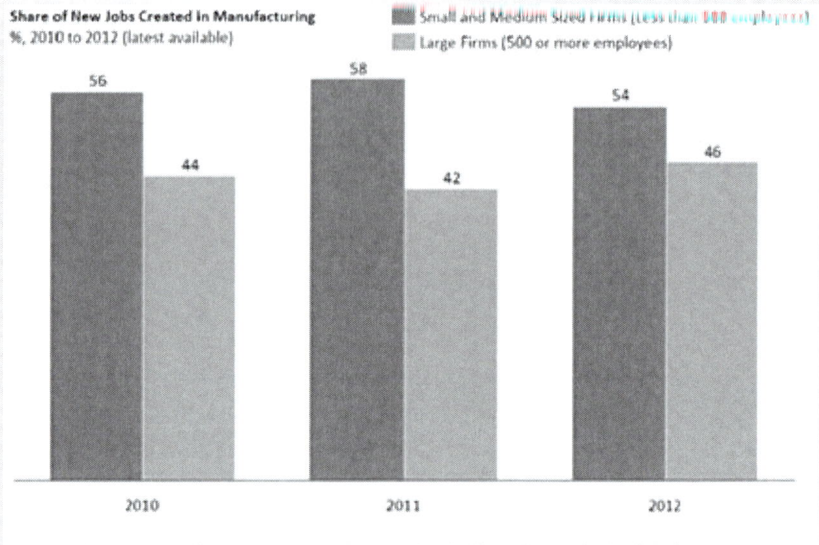

Source: Census Bureau's Business Dynamics Statistics.

The Challenge and Reward of Supply Chain Innovation

To stay at the forefront of global production, the U.S. manufacturing sector has stepped up its investment in research and development. U.S. manufacturing R&D, three-quarters of total U.S. private sector R&D, is at an all-time high of nearly $200 billion in 2013.[16] Manufacturers across the United States are investing in cutting-edge processes, advanced materials like carbon fiber and lightweight metals, new sensors and digital controls, and many more new technologies that have the potential to transform U.S. manufacturing.

New technologies such as these contribute to the nation's overall competitiveness only to the extent that they spread throughout U.S. supply chains. Dense networks of small manufacturers in the nation's supply chains are vital to the process of taking a product from concept to market, and the exchange of manufacturing know-how across suppliers is essential for the diffusion of these new technologies that give U.S. manufacturing its cutting edge.[17]

However, small manufacturers face barriers to innovation, with fewer resources, capabilities, internal engineering staff, and financial wherewithal to invest in research and development or to adopt new, potentially risky technologies, no matter how transformative. Economists often delineate three stages in the process of technical change: invention, the first appearance of a new idea; commercialization, the first practical application of the idea; and diffusion, the widespread adoption of the idea. Small suppliers face obstacles in each of these stages.[18]

Invention. Most private-sector R&D is performed by large manufacturers – very few small manufacturers invest in research and development internally which inhibits their ability to invent. Two-thirds of all manufacturing R&D is performed by just the 2 percent of manufacturing companies with more than 500 employees. In fact, 39 percent of all manufacturing R&D is attributable to just 20 large manufacturers, such as Boeing and Dow.[19] Only 2 percent of small firms conduct R&D at all, compared to 14 percent of large firms.[20]

At the same time, the creativity and agility of small, research-intensive manufacturers can be a source of innovation for the larger companies they supply, particularly as large manufacturers look increasingly to R&D partnerships for new ideas and breakthroughs. Indeed, some small manufacturers have made nimble innovation central to their business model. While overall few small firms conduct their own R&D, among those that did, these small businesses had on average 13 percent of their workforce in R&D, compared to only 6 percent at large firms conducting R&D.[21]

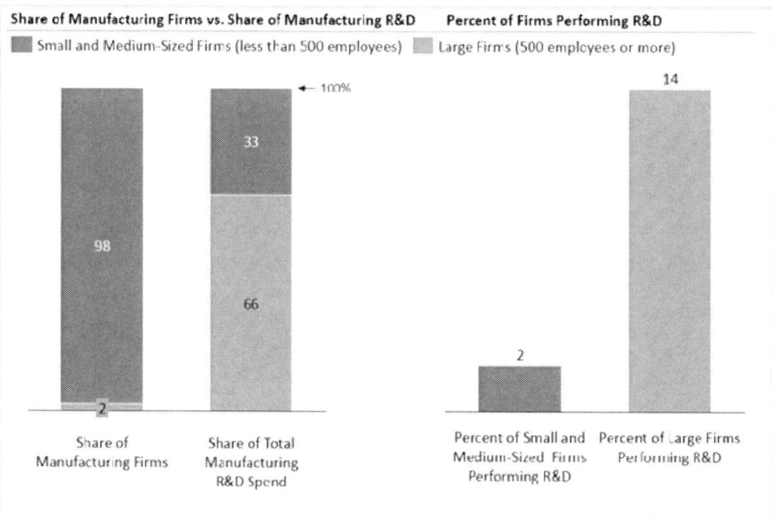

Source: National Science Foundation Business R&D and Innovation Survey (2005 & 2011); Census Bureau Business Dynamics Statistics (2012).

Commercialization. Even when small manufacturers do invest in R&D and this yields an invention, small manufacturers often have trouble finding the capital and connections needed for commercialization.[22] Small manufacturers' lower margins and limited access to capital hampers their investment in new technologies and the extent to which they take risks on technologies that have yet to be widely tested and adopted.[23]

Even with a relatively well-tested technology, upgrading a supplier's tooling and capabilities to meet the needs of a prospective customer can be risky. A supplier has to pay for the new equipment required to start producing a new part but will only get paid when that part is finally being supplied — months and sometimes years later. And if a supplier serves several large manufacturers, the large firms may lack incentive to invest in upgrading the supplier's capabilities if that supplier may also use those capabilities to serve a competitor.

Tighter research and technology linkages between large firms, suppliers, and other small manufacturers in the supply chain can enhance the diffusion of new technologies and lead to productive exchange of knowledge and capabilities. As the fixed cost of prototypes and digital design for smaller firms decreases, the potential for agile small manufacturers to develop new technologies, either on their own or in partnership with larger firms, increases. For example, GE, which re-shored its appliance manufacturing to take innovative products to market more rapidly, is partnering with Local Motors to "crowd source" ideas for new appliances and improvements to existing appliances and to develop prototypes to test these ideas more quickly and efficiently. 24

Diffusion. Small manufacturers are often slow to adopt new technologies. For example, a recent survey found that less than 60 percent of small manufacturers were experimenting in any way with 3-D printing, a potentially transformative technology that is especially beneficial for small companies due to its flexibility. In contrast, over 75 percent of large firms were using the technology.[25]

Since most R&D is performed by large firms, small manufacturers are usually in the position of needing to adopt a technology that was invented and tested outside their shop. Yet, small manufacturers employ fewer of the engineering and design staff needed to validate a new technology and update processes and products to incorporate it into production. For example, successful adoption of IT-based technology often requires complementary investments in equipment, shop-floor organization, marketing, and training – a scale and scope of expertise that challenges the capabilities of many small firms. [26,27] Small suppliers may also be unaware of new technologies because they lack the resources to track technological developments.[28]

Finally, industry dynamics can hinder investment by small firms. A small firm wishing to adopt new technologies may find it difficult to compete in the near term with rivals intent on competing by cutting corners rather than innovating. While some firms survive based on perennial cost-cutting of this type (such as using lower-quality materials or reducing investment in equipment and training), other small firms in the same industry wishing to invest in innovation can find it difficult to match the prices offered by these competitors. This situation can create a race to the bottom, preventing an entire industry from harnessing technological advances.[29]

In large part because of the significant challenges they face in inventing or adopting new technologies, small manufacturers are significantly less productive than their larger counterparts. Small manufacturing establishments are less than 60 percent as productive as their larger counterparts.[30] Small manufacturers' barriers to innovation ultimately affect all firms in the supply chain. The weaknesses of suppliers accumulate to adversely affect the lead firm, its final products, and its ability to introduce new products to market.

Today, America's small manufacturers, which undergird U.S. supply chains, are once again on the rise. But more needs to be done to rebuild their capabilities and tighten the linkages between firms that can accelerate products to market and spur fruitful cross-pollination of innovation and expertise across firms.

REINVESTING IN OUR NATION'S SUPPLY CHAINS AND SMALL MANUFACTURERS

The vitality of the nation's supply chains and the ability of America's small manufacturers to increase their capability with the latest technologies are important to U.S. competitiveness for jobs and investment in manufacturing, broadly. And the strength of its supply chains in many cases determines the ability of an individual manufacturer – even a large firm – to bring products successfully to market. As a result, reinvesting in America's small manufacturers, boosting their technological capabilities, and building tighter linkages between firms should be a public and a private sector priority.

Industry leaders, in particular, have a key role to play in strengthening U.S. supply chains by helping small manufacturers accelerate the adoption of new technologies. Through supplier mentoring and high-road business practices, large firms can expose small manufactures in their supply chains to new technologies while also providing them with greater security to take risks and realize a return on new technologies. In addition, small manufacturers and entire supply chains, by banding together in partnerships focused on technology exploration, can collectively lower the risk of investing in new technologies and coalesce around technology standards that can make collaboration across firms more efficient.

Undertaking these efforts represents a new strategic shift for many large firms which are coming to appreciate the increasing importance of innovative capacity in their supply chains rather than just high-volume, low-cost production. As McKinsey notes, "Many global supply chains are not equipped to cope with the world we are entering. Most were engineered, some brilliantly, to manage stable, high-volume production by capitalizing on labor-arbitrage opportunities available in China and other low-cost countries."[31] Absent a reinvestment in the innovative capacity of their suppliers and tighter linkages and coordination across the supply chain, many large manufacturers may struggle to keep up with the accelerating pace of innovation and consumer demand for rapid product cycles.

Evidence suggests that the following practices help firms promote supplier innovation, especially among small businesses in their supply chains:[32]

1. Offer suppliers assurance that they will receive a return on investments they make in new technologies and in upgrading their capabilities.

Investments in new capabilities and technologies are inherently risky, especially for small suppliers with tight margins and limited access to capital. Informal assurance from larger partners that these investments are of sufficient value to receive a price premium and continued business can reduce the risk of the investment. And guidance from larger manufacturers about emerging manufacturing technologies, especially those that may require collaborative changes in design between supplier and lead manufacturer, can help suppliers identify those technologies that provide the greatest value.

Bruno Independent Living Aids, a Wisconsin maker of accessibility solutions such as stair lifts, was able to convince a key supplier to make an investment in technology to solve its production needs through informal assurances. A few years ago, Bruno needed to find a way to inhibit rust on products used in outdoor environments. Bruno's painting supplier, Ad-Tech

Industries, invested in a new "E-coat" paint process that allows paint to adhere to metal very tightly using an electrostatic charge, thus reducing rust. The equipment cost over one million dollars, a large investment for a small supplier.

However, because Bruno was willing to make an informal assurance that it would buy capacity on the new machine at a reasonable price, Ad-Tech's risk in making the investment fell significantly. As a result, Bruno has a solution to its quality problem, and Ad-Tech has a new capability that benefits other customers as well, thus strengthening the whole manufacturing ecosystem.[33]

Similarly, Atlantic Tool and Die in Strongsville, Ohio had a single small factory when it began supplying small stamped parts to Honda in 1988. Today, Atlantic Tool and Die credits its partnership with Honda for enabling it to grow to six plants in the United States and abroad. In particular, Honda's technical assistance on new technologies and commitment to future business enabled Atlantic Tool and Die to invest in robotic welding that made it attractive to new customers while benefiting Honda through improved quality and economies of scale. The skill that Atlantic gained enabled the firm to invest, early on, in putting sensors in its dies, a technology that reduced stamping errors by more than 90 percent.[34]

2. Promote information-sharing and make changes in their own operations as a result of supplier suggestions.

Communication with suppliers may reveal actions by lead firms that turn out to be costly for suppliers – costs that ultimately may be passed on to the buying firm. For example, large companies often undertake policies such as frequent schedule changes or late payments that impose far greater costs on their suppliers than they realize. Other firms may allow suppliers to speak only to purchasing agents, in an attempt to preserve their bargaining power. However, blocking communication with engineers may mean that they miss out on suppliers' best ideas and joint problem-solving that would benefit both firms.

An example from Itron, a maker of gas and electric meters, shows the value of joint problem-solving. A new Itron product had a terminal in which the contact area was coated with silver and cadmium, both very expensive metals. Combining their expertise, Itron and the supplier were able to reduce the area needed for contact, saving over a third of the cost of the component. Itron's practice of frequent communication and joint problem-solving with its suppliers is a growing competitive strength for the company. 35 These benefits of coordination among firms are magnified in the case of a breakthrough technology, such as a new material that requires a different production or joining process.

3. Use a "Total Cost of Ownership" approach in making purchasing decisions.

Innovation is greatly facilitated if lead firms choose suppliers not on the narrow basis of low piece prices, but instead use a "Total Cost of Ownership" (TCO) approach, which considers the many kinds of costs to the buyer associated with the acquisition, transportation, and storage of products within the supply chain.[36] These costs include shipping costs, shipping time, storage costs and risk of damage or obsolescence of inventory, quality costs, travel costs for supplier oversight, currency fluctuations, intellectual property risk, financing costs, and risk of interruptions to the supply chain.[37]

Consideration of this broader set of costs encourages suppliers to innovate to find new ways to maximize value for the supply chain as a whole. For example, a supplier of advanced fuel systems was pursuing a large multi-year contract with one of the Detroit automakers. The automaker's purchasing department demanded a price reduction, noting that a competing supplier had submitted a bid at a significantly lower price per piece. The first supplier's patented catalyst technology used a wash coat technique to place precisely the precious metals used in its system, resulting in reduced use of the metals, lower cost of operation, and more effective oxidation of hydrocarbons and carbon monoxide. After the supplier's marketing team quantified and documented the improvements they offered, the automaker agreed to pay a premium compared to the lower-priced bidder. Even with the premium, the automaker's costs were lower than with the competing supplier because of the advanced technology.[38]

Research shows that practices such as those adopted by Itron, Bruno Independent Living Aids, and Honda – of involving suppliers early in product design, providing technical assistance, engaging in frequent communication, and being willing to share costs – are highly correlated with supplier willingness to invest in new technology.[39]

Unfortunately, these beneficial technology collaborations and efforts by larger companies to help supply chains upgrade their capabilities are too infrequent.[40] In some cases, large manufacturers will underinvest in their suppliers fearing that improvements in suppliers' capabilities might also serve their competitors.

Supply Chain Innovation in the Automobile Industry

Introducing breakthrough technologies benefits greatly from coordination among firms, including suppliers that can improvise, do new things, and understand the whole as full partners in innovation. One example is increasing the energy efficiency of cars by using lightweight materials: a ten percent reduction in a vehicle's weight can lead to a six to eight percent reduction in fuel use.

To introduce new, sophisticated light-weight metals into the production process requires combining the expertise of large automakers and their suppliers.

For example, "dual-phase" steel becomes strong only if it is stretched, so a lighter-weight part will fail unless ridges are included in the design to ensure that all areas of the part are stretched as it is stamped. In contrast to traditional steel, whose properties are well understood, designing the nature of the ridges required is still more of an art, and requires communication between the small tool and die shops, which have the knowledge of what kinds of ridges are needed, and the large automakers who want to use the new materials in their cars.

Here, industry learning overall can be accelerated by encouraging greater collaboration among automakers and their suppliers, and adopting supplier contracts that allow for experimentation by recognizing the reality of occasional failures. One example of collaboration is underway at the Center for Automotive Research (CAR) in Ann Arbor, Michigan. CAR leads the Coalition for Automotive Lightweighting Materials (CALM)—a consortium of 30 suppliers and small manufacturers that work together with automakers to identify barriers to the introduction of lightweight materials, and develop solutions enabling their use.

In other cases, a near-term focus on getting the lowest price for a particular component in the supply chain may obscure the long-term productivity benefits of encouraging a supplier's investments in new technologies, or preserving a supplier's margins for reinvestment in new capabilities.[41] And forming and sustaining cross-industry technology collaborations between small businesses is not easy – especially when many small manufacturers lack sufficient engineering and research personnel to easily participate in these collaborations.

The public sector can build on private-sector efforts such as those described above, acting as a catalyst for technology acceleration to benefit the supply chain as a whole. Government can help small manufacturers update their technology, processes, and capabilities, and help small manufacturers access state of the art research and equipment that normally would be out of reach for any one small company. The public sector can also play a catalytic role, by convening supply chain consortia to develop new technologies, and highlight best practices in customer-supplier relationships.

Promising initiatives along these lines are already underway:[42]

The Hollings Manufacturing Extension Partnership (MEP), created in 1988 and part of the Commerce Department's National Institute of Standards and Technology (NIST), was designed to overcome some of these challenges by providing small businesses access to management and technological expertise in 60 centers across the country. Every day, its more than 1,200 experts consult with manufacturers to help them improve their processes and identify opportunities to adopt new technologies or take new products to market. Today, over 30,000 manufacturers benefit from this expertise and network.

Recognizing the growing importance of supply chains and realizing that stronger connections between firms can help small manufacturers upgrade their capabilities, the Manufacturing Extension Partnership is expanding the successful supply chain and technology acceleration services for small manufacturers piloted over the last few years. These services include:

- *Supplier Improvement and Supply Chain Optimization* – MEP's Supplier Improvement program helps individual small manufacturers identify process improvements and technology upgrades to help them better compete within supply chains. And working across an entire supply chain, MEP's Supply Chain Optimization Program helps manufacturers build dynamic supply chains by assessing total cost of ownership, building deeper supplier communication, and helping suppliers upgrade their quality.[43]
- *Supplier Scouting and Business-to-Business Networks* – These programs help manufacturers find a U.S. supplier either through MEP's network of 30,000 manufacturers across the country or through new regional business-to-business network projects and manufacturing exchanges that directly link buyers to suppliers who have the right capabilities.
- *Supply Chain Technology Acceleration* – MEP helps small manufacturers within specific supply chains identify technologies, acquire technology-driven market intelligence, and mitigate potential

risk of new technologies by modeling their application to small manufacturers' existing processes.

In addition, federal agencies such as the Departments of Defense and Energy have begun to partner with firms of all sizes to strengthen supply chains in manufacturing industries important to their missions.

The Department of Defense Office of Economic Adjustment (OEA) supports state and local governments in responding to changes in defense programs that affect communities, including base closures or expansions, and incompatibilities between military operations and local development.[44] OEA's Defense Industry Adjustment (DIA) program works with supply chains, seeking to help its customers improve their resiliency in their respective defense industrial bases. While DIA works to help dislocated workers and impacted firms adjust to demand shocks from changes in defense spending, another major component of the DIA program is helping communities understand the economic clusters that comprise their local, regional, or state defense industrial base. In addition, DIA helps these communities understand how their industrial base interacts with other states or regions.

One recent DIA project is the Virginia DoD Procurement Economic Impact Evaluation Model.[45] The model is a supply-chain mapping of DoD contract awards, including sales and employment impacts, available for public use. It provides state-, county-, and local-level details about current and projected economic impacts of DoD contract spending. Similar projects are underway in Colorado, southern California, and New Jersey.

The Department of Energy (DoE) maintains 17 national labs in the United States. The labs engage in basic and applied technology R&D and provide a forum for the exchange of technology and ideas between regional firms, universities, and economic development intermediaries.[46] The New Mexico Small Business Assistance (NMSBA) program helps small businesses in the area by providing access to experts at the local Los Alamos National Laboratory and Sandia National Laboratories.[47] Technical assistance is funded by the state and provided to businesses free-of-charge, but access is competitive. To help small businesses compete for funding, the NMSBA created a national lab voucher program that since 2000 has helped over 1,000 small businesses gain access to the Los Alamos and Sandia labs. The state government provides the funding for the vouchers through a partnership with the NMSBA.

Ten projects developed by small businesses in New Mexico leveraging technical expertise and assistance provided by the Los Alamos and Sandia labs were recognized in 2012 for their innovation.[48] Technologies of Santa Fe, a small manufacturer, worked with the labs to develop a solar thermal ice-making product to deliver life-saving vaccines throughout the world in storage containers that use solar energy to maintain low temperatures.

The SupplierPay Initiative, a partnership between the federal government and the private sector, has resulted in nearly 50 companies signing a pledge to improve payment terms to their suppliers.[49] Because small suppliers are often subject to more stringent credit restraints than larger companies, they have difficulty securing working capital when buyers lengthen payment terms. This, in turn, passes costs back onto the buyers in the form of higher piece prices or reduced innovation. Faster payment terms free up working capital for small suppliers to use for investments in worker training orinnovation.[50] Many of these companies also play a prominent role in the American Supplier Initiative, often participating as corporate buyers in a series of nationwide events designed to engage small business suppliers in corporate supply chains.

Restoring the vitality of America's supply chains and the small manufacturers at their core is essential to the long-term competitiveness of U.S. manufacturing. The Administration's focus on reinvesting in America's supply chains is just one example of how joint public-private sector efforts can strengthen the foundation of these key elements of the U.S. economy.

Now is the time for additional action to build on these efforts and strengthen our manufacturing base. As discussed above, market-based incentives by themselves are inadequate to ensure sufficient investment in supply chains. There are ample opportunities for government at all levels to partner with the private sector to promote supply chain innovation and thus foster a continued manufacturing renaissance, both now and into the future.

APPENDIX: AN INTERNATIONAL PERSPECTIVE

Governments around the world recognize the importance of fostering innovation throughout the supply chain. Below are examples of these efforts from around the globe.

Germany

German firms have a diverse set of resources at their disposal. Both the firms and the government contribute to the development and maintenance of these relationships.[51] Firms invest heavily in the capabilities of their workers, many of whom also participate in apprenticeship programs sponsored by the government. When firms consider bringing a new idea to market, they often rely on collaboration with their suppliers to develop new technologies and work through the initial phases of scale-up together.[52]

The industrial ecosystem in Germany involves many players – lead firms and suppliers, trade associations, unions, industrial collective research consortia, industrial research centers and associations, Fraunhofer Institutes, university-industry collaboratives, technical advisory committees, and others – who all work together to create an environment that supports innovation.[53] These collaborations are numerous and are not a new phenomenon—a survey of 744 industrial collective research projects in German between 2003 and 2005 identified almost 300 organizations involved in these projects.[54] The German government provides some limited funding through these organizations, but the convening power of government is one of the greatest benefits to all firms from these arrangements.[55] The government funding is used to incentivize collaborations; for example, the government provided 40 million Euros (matched by private industry) to the photovoltaic cluster Solarvalley Mitteldeutschland to match the same funding from industry for a five-year project on grid parity that brought together firms, universities, and research institutes.[56] They also provide innovation vouchers to firms to help catalyze innovation and connect them with researchers at universities or other firms.[57]

United Kingdom

In the United Kingdom, the Manufacturing Advisory Service (MAS), through nine regional centers, provides multiple supports to SMEs. Services include a helpline providing technical assistance, consulting services on process improvement, and networking and training events including best-practices factory visits.[58] These services complement existing supplier mentorship programs provided by lead firms.[59]

The UK has also created a Technology Strategy Board (TSB) with a goal of increasing supply chain R&D collaboration. The TSB also facilitates supply chain partnership opportunities through its Collaborative R&D calls and Feasibility Studies where companies can test new ideas and get customer feedback. Many of these collaborations take place at seven High Value Manufacturing Catapult (HVM Catapult) centers. HVM Catapult provides companies and suppliers open-access equipment and technical assistance. An example of a technological innovation that came out of a HVM Catapult center was the MEGA-FLUTE tooling technology, which has the potential to revolutionize milling cutter technology in the aerospace industry, leading to significant reductions in cost and capital expenditures.[60]

Canada[61]

Canada's Industrial Research Assistance Program (IRAP) helps SMEs build their innovation and new product development capacity by providing technology assistance through a network of 220 Industrial Technology Advisors (ITAs) located in 90 communities throughout Canada.[62] The ITAs connect SMEs with Canadian universities and national laboratories so that the institutions can share emerging technologies and knowledge. IRAP also directly funds R&D and innovation at SMEs.

As in Germany, the Canadian government provides innovation vouchers to firms and provides financial support to SMEs to hire recent college graduates. In some cases, the newly-minted employees with technical degrees are partially supervised by IRAP personnel. A key issue in having small firms use the innovation vouchers is defining an appropriate problem – one which can be solved by the labs and whose solution is also beneficial to firms. The hands-on involvement by experienced IRAP personnel, plus the proximity to the operation of the recent graduate, helps firms to define a problem for which they could productively use an innovation voucher or other program.

Productivity Alberta, started by the government of Alberta in 2008, has assigned MBA students to SMEs to identify and help solve innovation, technical, and scientific challenges by leveraging resources at their graduate schools.[63]

End Notes

[1] Economics and Statistics Administration analysis using data from the Bureau of Labor Statistics' Current Employment Statistics (CES). CES data may be accessed online at: http://www.bls.gov/ces/.

[2] Economics and Statistics Administration analysis using data from the Census Bureau's Business Dynamics Statistics (BDS). BDS data may be accessed online at: http://www.census.gov/ces/dataproducts/bds/.

[3] See Berger, Suzanne, Making in America: From Innovation to Market, MIT Task Force on Production in the Innovation Economy, The MIT Press, 2013; Gary P. Pisano and Willy C. Shih, Producing Prosperity: Why America Needs a Manufacturing Renaissance, Harvard Business Review Press, 2012.

[4] Economics and Statistics Administration analysis using data from the National Science Foundation's 2011 Business R&D and Innovation Survey (BRDIS) and the Census Bureau's 2011 Statistics of U.S. Businesses (SUSB). BRDIS data may be accessed online at: http://www.nsf.gov/statistics/srvyindustry/about/brdis/. SUSB data may be accessed online at: http://www.census.gov/econ/susb/.

[5] National Institute of Standards and Technology analysis using data from the Census Bureau's 2007 Economic Census. Economic Census data may be accessed online at: http://www.census.gov/econ/census/.

[6] Economics and Statistics Administration analysis using data from the Census Bureau's 2012 Economic Census. For this calculation, purchased inputs includes: materials, services and expenses, contract work, and energy.

[7] Dedrick, Jason, Kenneth L. Kraemer, and Greg Linden, "Who Profits from Innovation in Global Value Chains? A Study of the iPod and notebook PCs," Personal Computing Industry Center, UC Irvine, May 2008, available at: http://web.mit.edu/is08/pdf/Dedrick Kraemer Linden.pdf (last accessed March 2015); A.D. Chandler Jr., The Visible Hand: The Managerial Revolution in American Business, Harvard University Press, 1977.

[8] See The Henry Ford webpage, History of the River Rouge, available at: - http://www.thehenryford.org/rouge/historyofrouge.aspx (last accessed January 2015).

[9] Klier, Thomas and James Rubenstein, Who Really Made your car?: Restructuring and Geographic Change in the Auto Industry, Upjohn Press, 2008.

[10] U.S. Census Bureau's Economic Census, 2012

[11] The White House, "The Resurgence of the American Automotive Industry," Office of the Press Secretary, June 2011, available at: http://www.whitehouse.gov/sites/default/files/uploads/auto report 06 01 11.pdf (last accessed January 2015).

[12] Ibid.

[13] Economics and Statistics Administration analysis using data from the Census Bureau's Business Dynamics Statistics (BDS).

[14] Economics and Statistics Administration analysis using data from the Census Bureau's Business Dynamics Statistics (BDS). BDS shows that from 1980 to 2012, 52 percent of new manufacturing jobs were created by firms with fewer than 500 employees. During that time, only 46 percent of the job losses were from these firms, so they contributed positively to job creation over this time.

[15] Ibid.

[16] Dollar values are inflation-adjusted. Data from Bureau of Economic Analysis, Table 5.6.6. Available at: http://www.bea.gov/iTable/ iTable. cfm?reqid=9&step=3&isuri=1&903=332#reqid=9&step=3&isuri=1&903=332.

[17] Berger, Suzanne, Making in America: From Innovation to Market, MIT Task Force on Production in the Innovation Economy, The MIT Press, 2013; Gary P. Pisano and Willy C. Shih, Producing Prosperity: Why America Needs a Manufacturing Renaissance, Harvard Business Review Press, 2012.

[18] One might argue that a lack of innovation by small firms is not problematic, i.e., that it makes sense for lead firms to specialize in innovation, while suppliers focus on production. However, increasing evidence suggests that this division of labor may be suboptimal. First, there are significant spillovers from production to product innovation; when product engineers are not near the production site, there may be fewer product innovations. Second, lack of process knowledge means that suppliers are less likely to generate or adopt innovative processes. See Pisano and Shih.

[19] Economics and Statistics Administration analysis using data from National Science Foundation's 2005 Business R&D and Innovation Survey.

[20] Economics and Statistics Administration analysis using data from the National Science Foundation's 2011 Business R&D and Innovation Survey and the Census Bureau's 2011 SUSB.

[21] Economics and Statistics Administration analysis using data from the Census Bureau's 2011 BRDIS.

[22] Advanced Manufacturing Partnership report, "Accelerating U.S. Advanced Manufacturing," President's Council of Advisors on Science and Technology, Executive Office of the President, October 2014, available at: http://www.whitehouse.gov/sites/default/files/ microsites/ostp/PCAST/amp20 report final.pdf (last accessed March 2015).

[23] See Helper, Nicholson, and Noonan. "The Economic Benefits of Reducing Supplier Working Capital Costs," available at: http://www.esa.doc.gov/ sites/default/files/supplierpayv25.pdf (last accessed March 2015). 24Gabrielle Karol, "Local Motors Bringing Crowdsourced Innovation to GE," Fox Business, March 18, 2014, available at http://smallbusiness. foxbusiness.com/technology-web/2014/03/18/local-motors-bringing-crowdsourced-innovation-to-ge/.

[25] PWC, "3D printing and the new shape of industrial manufacturing," June 2014, available at: https://www.pwc.se/sv SE/se/verkstad/assets/3d-printing-and-the-new-shape-of-industrial-manufacturing.pdf (last accessed March 2015).

[26] Kelley, Maryellen and Susan Helper, "Firm Size and Capabilities, Regional Agglomeration, and the Adoption of New Technology," Economics of Innovation and New Technology, Vol. 8, Issue 1-2, 1999, available at: http://www.tandfonline.com/doi/abs/ 10.1080/ 10438599900000005?journalCode=gein20#.VP3w9 zF98E (last accessed March 2015); Daniel D. Luria, "Why Markets Tolerate Mediocre Manufacturing," Challenge, July-August 1996, available at: http://business.highbeam.com/137644/article-1G1-18544973/why-markets-tolerate-mediocremanufacturing (last accessed March 2015).

[27] Brynjolfsson, Erik and Lorin M. Hitt, "Computing Productivity: Firm-Level Evidence," MIT Sloan Working Paper No. 4210-01, June 2003, available at: http://papers.ssrn.com/ sol3/papers.cfm?abstract id=290325 (last accessed March 2015); Bartel, Ann, Casey Ichniowski, and Kathryn Shaw, "How Does Information Technology Affect Productivity? Plant-Level Comparisons of Product Innovation, Process Improvement, and Worker Skills," Quarterly Journal of Economics, November 2007, available at: http://www.ppge.ufrgs.br/ GIACOMO/arquivos/ecop 137/ bartel-ichniowski-shaw-2007.pdf (last accessed March 2015).

[28] Ibid.

[29] Daron Acemoglu, "Good Jobs versus Bad Jobs," Journal of Labor Economics 19 (2001), 1-21; Michael Porter, Competitive Advantage: Creating and Sustaining Superior Performance. New York: Simon and Schuster, 1980.

[30] National Institute of Standards and Technology analysis using data from the Census Bureau's 2007 Economic Census. This productivity differential is not explained by large and small firms competing in different industries; these productivity differences persist even when controlling for narrow industries. In fact, industry effects explain only about 10 percent of productivity differentials. Within 4-digit NAICS codes, the most productive firms are, on average, three times as productive as the least productive firms. One determinant of these intra-industry differentials is firm size. See Syverson, Chad, "The Importance of Measuring Dispersion in Firm-Level Outcomes," IZA World of Labor, May 2014, available at: http://home.uchicago.edu/syverson/ Syverson WoL.pdf (last accessed January 2015).

[31] Malik, Yogesh; Alex Niemeyer; and Brian Ruwadi, "Building the supply chain of the future," McKinsey & Company, McKinsey Quarterly, January 2011, available at: http://www.mckinsey.com/insights/ operations/ building the supply chain of the future.

[32] A vast literature describes the potential benefits to lead firms of collaborative relations with suppliers. A review article finds significant benefits to trusting relationships in terms of reduced cost, defect rates, lead times, and increases in investment, responsiveness, and problem-solving behavior. Delbufalo, Emanuela, "Outcomes of inter-organizational trust in supply chain relationships: a systematic literature review and a meta-analysis of the empirical evidence", Supply Chain Management: An International Journal, Vol. 17 Issue 4 –2012, available at: http://dx.doi.org/10.1108/13598541211246549 (last accessed March 2015). Extensive study suggests that a key part of the success of firms such as Toyota and Honda was due to their investments in such relationships with their suppliers. Gibbons, Robert and Rebecca Henderson, "Relational Contracts and Organizational Capabilities," Organization Science, Volume 23, Issue 5, September-October 2012 Note that being a Toyota supplier was not "a cozy relationship," as one manager of a supplier company pointed out (see Susan Helper and Rebecca Henderson, "Management Practices, Relational Contracts, and the Decline of General Motors," Journal of Economic Perspectives, Vol. 28, No. 1, Winter 2014, available at: https://www.aeaweb.org/articles.php? doi= 10.1257/jep.28.1.49). Toyota pushed its suppliers very hard to reduce costs and avoid defects; it reduced the market share of suppliers who did not meet these strict goals; and it exited the relationship completely if improvement was not forthcoming.

[33] Bruno Independent Living Aids, interview and plant visit), Oconomowoc, WI, February 27, 2015.

[34] Stevens, Merieke, John Paul MacDuffie, Susan Helper, "Reorienting and recalibrating inter-organizational relationships: Strategies for achieving optimal trust," Organization Studies, forthcoming.

[35] Itron, interview and plant visit, West Union, SC, February 24, 2015

[36] NIST MEP, "U.S. Supply Chains: Insights into the Challenges and a Foundational Roadmap toward Global Competitiveness," U.S. Department of Commerce, available at: http://mepsupplychain.org/wpcontent/uploads/2013/08/SupplyChainWhitepaper.pdf (March 2015).

[37] See Department of Commerce's Assess Costs Everywhere (ACE) tool at: http://acetool. commerce.gov/ and Terry Weiner, "The Importance of Total Cost of Ownership (TCO)," CMTC Manufacturing Blog, January 21, 2014, available at: http://www.cmtc.com/blog/ bid/156131/The-Importance-of-Total-Cost-of-Ownership-TCO (last accessed January 2015).

[38] Kotler Marketing Group, private communication, March 5, 2015.

[39] Zhang, Chun, John W. Henke Jr., and Sridhar Viswanathan, "Reciprocity between buyer cost sharing and supplier technology sharing," Journal of Operations Management, 27, 2009, available at: http://www.ppi1.com/wpcontent/uploads/2012/09/JOM-Supplier-Innovation-Nov-Dec-2009.pdf (last accessed March 2015.)

[40] For example, the technique used by Itron in the example above, called "value analysis/value engineering" is not widely adopted in the US despite strong evidence of its effectiveness. A 2011 survey of auto suppliers found that less than one-third had adopted the technique. Jenny Kuan and Susan Helper, "What Goes on Under the Hood? How Engineers Innovate in the Automotive Supply Chain," in Engineering in a Global Economy, Richard Freeman and Hal Salzman, eds, NBER, forthcoming.

[41] For evidence of "short-termism" generally, see "Report of the Commission on Inclusive Prosperity," Center for American Progress, January 2015, available at: https://cdn.americanprogress.org/wpcontent/uploads/2015/ 01/IPC-PDF-full.pdf (last accessed March 2015).

[42] The appendix provides examples of programs in other nations that may be instructive for the U.S.

[43] These new services build on Accelerate, a Wisconsin Manufacturing Extension Partnership (WMEP) program that is used by over 400 suppliers across many different industries. The goal of Accelerate is to assist manufacturers to reduce their Manufacturing Critical-path Time (MCT). MCT is defined as the time between when a customer places an order to receipt of the first piece of that order. Reducing lead times is important because longer lead times cause firms to hold more inventory, often adding 20-30 percent to product costs. The U.S. Defense Logistics Agency recently reviewed 195 Department of Defense suppliers that had received training, and found that each company saw an average reduction in inventory of $132,000, as well as a $98,000 reduction in internal quality-related waste, and a $75,000 reduction in other direct operational costs. In addition, these suppliers experienced improvements in their on-time deliveries and their quality, as measured by their customers. Orders saw quicker turnaround, and MCT was cut by 50 percent.

[44] For more information, see the OEA's website, available at: http://www.oea.gov/about/oea.

[45] For more information, see: http://www.chmuraecon.com/blog/2014/may/ virginia-department-of-defense-dodprocurement-economic-impact-evaluation-model/.

[46] Andes, Scott, Mark Muro, and Matthew Stepp, "Going Local: Connecting the National Labs to their Regions for Innovation and Growth," Brookings, "The information Technology & Innovation Foundation, and the Center for Clean Energy Innovation," September 2014, available at: http://www.brookings.edu/~/media/Research/Files/Reports/2014/09/10-national-labs/BMPP DOE Brief.pdf?la=en (last accessed March 2015).

[47] New Mexico Small Business Administration brochure, available online at: http://www.nmsbaprogram.org/userfiles/2014 NMSBA ProspectBro January 2015 FNL.pdf.

[48] "Los Alamos, Sandia National Laboratories recognize New Mexico small businesses for innovation," New Mexico Small Business Administration, May 1, 2012, available at: http://www.nmsbaprogram.org/news/detail/22 (last accessed March 2015).

[49] The White House, Office of the Press Secretary, "FACT SHEET: President Obama's SupplierPay Initiative Expands; 21 Additional Companies Pledge to Strengthen America's Small Businesses," November 17, 2014, available at: http://www.whitehouse.gov/the-press-office/2014/11/17/fact-sheet-president-obama-s-supplierpay-initiativeexpands-21-additiona (last accessed March 2015).

[50] See Helper, Nicholson, and Noonan. "The Economic Benefits of Reducing Supplier Working Capital Costs," available at: http://www.esa.doc.gov/sites/default/files/supplierpayv25.pdf (last accessed March 2015).

[51] Berger. Also, several German states co-finance the placement of recent Ph.D. graduates with SME manufacturers. In Brandenburg state's program, the state covers 50 percent of the cost for an SME to employ a recent Ph.D. graduate for up to two years according to in person interviews conducted by Stephen Ezell and Robert D. Atkinson. See "International Benchmarking of Countries' Policies and Programs Supporting SME Manufacturers," Information Technology and Innovation Foundation, September 14, 2011, available at: http://www.itif.org/ publications/international-benchmarking-countries-policies-and-programs-supporting-smemanufacturers (last accessed March 2015).

[52] Berger.

[53] Fraunhofer is Europe's largest application-oriented research organization. Its 66 institutes and research units are located throughout Germany. See more at: http://www.fraunhofer.de/en/about-fraunhofer.html.

[54] Berger.

[55] Ibid.

[56] Ibid.

[57] Ezell and Atkinson.

[58] Ibid.

[59] United Kingdom Department for Business and Innovation Skills, Strengthening UK Supply Chains, January 2014, available at: http://www.mbsportal.bl.uk/taster/subjareas/techinnov/bis/15994914 515 strengthening supply.pdf.

[60] Ibid.

[61] Ezell and Atkinson.

[62] See also the National Research Council of Canada, "IRAP: Mandate, Mission, Values, Program Goals and Strategic Objectives," available at: http://www.nrc-cnrc.gc.ca/eng/irap/index.html.

[63] Ezell and Atkinson interview with Lori Schmidt, Senior Director, Productivity Alberta, July 6, 2011 as cited in footnote 58. In 2011, Productivity Alberta became Go Productivity, a private non-profit that helps Canadian companies identify and address gaps in productivity, maximize resources, and be more efficient. For more information see the Go Productivity website at: http://goproductivity.ca/.

In: Small Business Innovators
Editor: Mary Brown

ISBN: 978-1-63483-265-6

Chapter 3

INNOVATION ACCELERATORS: DEFINING CHARACTERISTICS AMONG STARTUP ASSISTANCE ORGANIZATIONS*

C. Scott Dempwolf, Jennifer Auer and Michelle D'Ippolito

ABBREVIATIONS

IP – Intellectual Property
IPO – Initial Public Offering
NBIA – The National Business Incubation Association
NESTA – National Endowment for Science Technology and the Arts
POCC – Proof-of-Concept Center
RIAN – Regional Innovation Acceleration Network
SBIR – Small Business Innovation Research program
STTR – Small Business Technology Transfer program
VC – Venture Capital
VDO – Venture Development Organization

* This is an edited, reformatted and augmented version of contracted report, issued by the U.S. Small Business Administration, Office of Advocacy, October 2014.

1. Executive Summary

The accelerator phenomenon has been cited nationally and internationally as a key contributor to the rate of business startup success. Accelerators select and invite a small group of entrepreneurs to startup boot camps, providing mentoring, resources, and, most important, industry connections during these programs. Successful graduates leave with the next stage of funding or revenue in hand, and all graduates leave with a small percentage of their company's equity ceded to the accelerator. Accelerators address the funding gap for startups and the information gap for would-be investors by acting as network brokers. They reduce search costs for angel funders and venture capitalists while creating a pipeline of vetted technologies for the market. By concentrating resources through seed funding, access to investment networks, and intensive mentoring, accelerators are able to identify "winning" ideas more quickly and help those startups grow.

Although successful examples of high growth startup companies are reported by accelerators like Y Combinator in Northern California (e.g., Dropbox, Reddit), accelerator advocates have encountered criticism for the lack of statistical data and metrics to validate this impression of success. Do accelerators produce viable companies? Increase innovation? Create jobs? Produce windfalls for their founders? There is debate concerning the accelerator model's potential, and resolution of this debate is partially hindered by the fact that thus far there has been no consensus in the literature regarding what defines an accelerator and how to distinguish the many types of accelerators from other startup assistance programs, such as business incubators.

To help bring clarity to the debate, this study creates a taxonomy of startup assistance organizations and provides a working definition of an innovation accelerator that departs from those found in the existing literature. Previous definitions have highlighted accelerators' services and focus on software applications as key characteristics of the definition. The proposed taxonomy distinguishes accelerators from other startup assistance organizations based on the organization's value proposition and business model, both of which are influenced significantly by the accelerator's technology focus and the founder's motivation for starting. Through this taxonomy, three categories of startup assistance organizations are identified: (1) incubators and venture development organizations, (2) proof-of-concept centers, and (3) accelerators. Accelerators are further subdivided into social accelerators, university accelerators, corporate accelerators, and innovation accelerators.

Further, this study discusses the metrics that should and could be measured according to the taxonomy and the definition proposed, and concludes that metrics designed to look for long-term outcomes provide the most appropriate assessment of accelerator performance. Nonetheless, the report questions the public value generated by innovation accelerators versus other subtypes, and demonstrates that if public resources are used to support accelerators, it is important to consider the accelerator subtype.

2. Introduction

Although they are a relatively new phenomenon, accelerators have garnered intense interest from the private and public sectors in the United States and abroad. The number of new accelerator programs has grown substantially since the launch of the popular accelerator Y Combinator in 2005.The phenomenon has inspired a wide range of events, such as the White House's endorsement of a Global Accelerator Network in 2010, and even a currently airing Amazon Instant Video series, *Betas*, depicting the trials of one fictional accelerator cohort.

Accelerators are programs that help entrepreneurs bring their technologies, ideas, or products into the marketplace and ideally lead entrepreneurs to develop viable businesses. Accelerators may appear to be very much like private-sector versions of business incubators, which have been in existence since the early 1960s. Accelerators and incubators work with startup businesses and provide a range of assistance and support services to these companies. Both prefer to work with for-profit startups with high-growth potential, and accelerators deal exclusively with these types of businesses. Both are focused on coaching and developing these startups so that they can secure early-stage "seed" funding. Despite their seemingly similar activities, accelerators are quite different from incubators. However, substantive differences have not been comprehensively documented in the accelerator literature to date.

This study compares accelerators and incubators, placing both within the larger context of an innovation ecosystem.[1] Such a comparison gives rise to a series of important questions that are due in part to the obvious appeal and rapid rise of accelerators. The questions that are central to discussions later in this report include the following:

1) What are accelerators?
2) Approximately how many currently exist?
3) Are accelerators and incubators the same thing?
4) Are there different types of accelerators, and if so, how do they differ?
5) Do accelerators have the potential to accelerate economic growth?
6) How can the performance and impact of accelerators be measured?

This research seeks to answer the first four questions and contribute to an ongoing dialogue surrounding the last two. The study also uses information from the first four questions to advance current debates concerning accelerators as a business model. For example, it is important to determine whether or not the accelerator model is applicable to multiple industries, such as advanced manufacturing. If not, does the business model embodied by accelerators only work for industries where the majority of existing accelerators have purportedly concentrated (i.e., software development and mobile device applications)? Is there a market failure that warrants a role for government as it relates to accelerators? Should the government assume a regulatory or incentivizing role for some accelerator subtypes?

The U.S. economy is increasingly dependent on innovation as a source of growth, and public programs actively support entrepreneurship in a variety of ways. Answering the questions stated above requires understanding the accelerator phenomenon from its roots. It calls for a grounded definition that accounts for the broader goals of innovation ecosystems. Without this base, it will not be possible to move to empirical studies that compare return on investment, the effects of different program durations, the relative outcomes when accelerators are employed in different industries, or the many other questions that policy should consider as it allocates attention across business-development strategies.

To develop a working definition and general understanding of accelerators, this report starts with a survey of existing literature (section 3). This section provides background about the available literature and identifies characteristics associated with accelerators, particularly in comparison to incubators. It also discusses the range of accelerator programs and organizations that have started over the past 10 years, and differences between accelerators found within the United States and abroad. Section 4 discusses the taxonomy that forms the center of the working definition of accelerators, including its methodology and framework. The taxonomy identifies important and defining differences between accelerator variants, and between accelerators and incubators. The section ends with a flowchart to facilitate classification.

Section 5 reviews the data sources and metrics identified in the literature review as they relate to defining accelerator success. This section focuses primarily on performance evaluations of accelerators and their startups, and ends with considerations for designing an evaluation system. Section 6 summarizes the results of the previous three sections to discuss the policy implications related to accelerators.

3. Literature Review

Current State of the Literature and Methodological Implications

What is known about accelerators is based on the small number of peer-reviewed articles and secondary media sources. The dearth of peer-reviewed literature can be attributed not only to the newness of the accelerator phenomenon—scholars have only had a few years to study them—but also to the fact that accelerators are generally privately held and funded, meaning they have no obligation to disclose information about their programs. Due to the limited number of peer-reviewed articles, it is currently necessary to rely on media sources or "gray" literature to define and characterize accelerators. Indeed, secondary media sources, including online articles, policy briefings, and interviews and correspondence with experts, are commonly referred to in the few academic articles that exist. A fuller description of the search strategy conducted for this literature review is provided in appendix 1.

On the whole, secondary media sources provide useful information to fill gaps in understanding about accelerators in the short term and potentially even in the longer term. However, the reliance on media-related sources in the scholarly literature has contributed to a lack of definitional clarity regarding accelerators as a distinct economic phenomenon. This situation makes it difficult to reliably synthesize findings on accelerator attributes, such as services, and ultimately, their performance. The information presented below highlights those articles that are aimed at delineating accelerators from incubators and more recent articles that have proposed definitions of accelerator programs. It also presents information on early, well-known accelerators and the spread of programs across the U.S. and abroad. The systematic definition and classification of accelerators and related entities—

the focus of this paper—is a necessary first step before truly reviewing the accelerator phenomenon and its outcomes.

Characterizing and Defining Accelerators

Accelerators are programs that help entrepreneurs bring their products into the marketplace. They typically operate by inviting a cohort of startup companies to work intensively on their technologies for a period of time. Early articles on the accelerator phenomenon defined them by the unique services they provide to entrepreneurs. For example, experts at the Kauffman Foundation explain that accelerators are organizations offering a suite of professional services, mentoring, and office space in a competitive program format (Fishback et al. 2007).

As the number of accelerator programs has grown substantially, scholars have since observed that a more precise definition is needed, especially to distinguish accelerators from business incubators. Both incubators and accelerators can be broadly characterized as groups of experienced businesspersons who provide nascent firms with advice, businesses services, financing on occasion, and often office space to help them develop and launch their businesses with greater success than if the startups had not received assistance (Bøllingtoft and Ulhoi 2005; Hoffman and Radojevich-Kelley 2012; Isabelle 2013). Yet, business incubators have been a popular means to support startups since at least the early 1980s (Allen and Rahman 1985).

The National Business Incubation Association (NBIA)—a membership organization for the incubator industry—derived a useful list of characteristics shared by incubators and accelerators. The analysis indicates that their differences sometimes lie in the nature, intensity, and duration of a characteristic, rather than its presence or absence in a program. For example, NBIA describes its members' startups as crossing numerous industries, ages, and experience levels, whereas it characterizes accelerators as being focused on Web-based technologies and their commensurate youthful and male-dominated founders. NBIA describes accelerator services as designed to quickly move startups from one stage to the next, while incubator services aim to move entrepreneurs toward self-sustaining, mature businesses. Importantly, incubators rarely invest directly in their startups, while accelerators typically provide funding in exchange for an equity stake in the startup's future profits. Table 1 presents other aspects of NBIA's comparison.

Table 1. Characteristics of incubators and accelerators

Characteristics	Incubators	Accelerators
Clients	All kinds, including science-based businesses (biotech, medical devices, nanotechnology, clean energy, etc.) and nontechnology; all ages and genders; includes those with previous experience in an industry or sector.	Web-based, mobile apps, social networking, gaming, cloud-based, software, etc.; firms that do not require significant immediate investment or proof of concept; primarily youthful, often male technology enthusiasts, gamers, and hackers.
Selection Process	Competitive selection, mostly from the local community.	Competitive selection of firms from wide regions or even nationally (or globally).
Terms of Assistance	1 to 5 or more years (33 months on average)	Generally 1- to 3-month boot camps
Services	Offers access to management and other consulting, specialized intellectual property and networks of experienced entrepreneurs; helps businesses mature to self-sustaining or high-growth stage; helps entrepreneurs round out skills, develop a management team, and, often, obtain external financing.	"Fast-test" validation of ideas; opportunities to create a functioning beta and find initial customers; linkage of entrepreneurs to business consulting and experienced entrepreneurs in the Web or mobile apps space; assistance in preparing pitches to try to obtain follow-up investment.
Investment	Usually does not have funds to invest directly in the company; more frequently than not, does not take equity.	Invests $18,000 to $25,000 in teams of co-founders; takes equity in every investee (usually 4 to 8 percent).

Source: Excepts from Atkins, D. 2011. What are the new seed or venture accelerators? Available at http://www.nbia.org/resource_library/review_archive/0611_01.php.

Other scholars and institutions (including NBIA) have highlighted differences in the characteristics of organizations hosting accelerator programs. From a policy perspective, the organizational distinctions among these entities are useful. Isabelle (2013), Atkins (2011), and Hoffman and Radojevich-Kelley (2012) spelled out several of these distinguishing characteristics. They found the typical characteristics associated with incubators to include the following:

- They are nonprofit organizations, frequently associated with universities.
- They provide office space at reasonable rates for the startups they support.
- They target local startups.
- They do not invest in the startups.

On the other hand, they found the following characteristics to be more typical of accelerators:

- They are for-profit organizations that receive equity in exchange for the provision of funding to the startups.
- They do not necessarily provide office space for the startups they support, but typically provide meeting space.
- They target regional, national, or even global startups.

The blurred identities of incubators and accelerators have partly been of interest because incubators are recognized by scholars and policy-makers as an established local economic development tool when, in fact, accelerators were popularized in the private sector. The role of accelerators in the larger innovation ecosystem, including among other publicly supported organizations, is not yet clear. For example, in its simplest form, innovation begins with basic scientific discovery, which leads to the development of new technologies and technology platforms from which a wide array of products may be invented. This process of innovation continues with the identification of funding for product development, testing, production, and marketing of those products for eventual sale in the marketplace—a process often referred to simply as "commercialization."[2] Startup assistance and business development organizations facilitate that process in various ways. Regardless of an accelerator's place in the overall ecosystem, the perceived success of accelerators in fostering new companies has led state governments to adopt variants of the model and provide funding to local organizations that help attract, nurture, and support venture-backed, high-growth startups (Sparks 2013).

The similar and sometimes overlapping characteristics of accelerators and incubators have led to some inconsistency in classification. Organizations may refer to themselves as being one type when their operations more closely reflect the other. In the absence of a standardized definition, scholars and practitioners have arrived at their own definitions, and each one is different. For example, several open-source data sources such as Seed-DB, maintain online listings of accelerators, while scholarly reports, including Miller and Bound (2011) and Christiansen (2009),[3] may provide accelerator lists based on the definition created in the report. Few accelerators are found across all lists.

This study cross-referenced six accelerator inventories to compile a list of over 80 accelerator organizations or programs that appear in at least two of the inventories. This process helped identify entities that are considered accelerators (rather than an incubator, for example) by more than one source. The accelerators, along with the location and founding date, are shown in appendix 2. The appendix indicates that accelerators operate not only in regions with well-known entrepreneur cultures (e.g., San Jose, New York, Austin), but also in regions that may be unexpected (e.g., Cincinnati, Provo). Many operate independently, but some well-established accelerators, such as Techstars, have branches in multiple cities. Table 2 shows the very few accelerators (or accelerator program sites) that were identified in all six lists.

Table 2. Frequently cited accelerators

Capital Factory	Austin TX
DreamIT Ventures	Philadelphia, PA location
Techstars	Boston, MA location
Techstars	Boulder, CO location
Y Combinator	Mountain View, CA

Outside the United States, accelerators are emerging throughout Canada, South America, Asia, the United Kingdom, and many European countries. According to the National Endowment for Science Technology and the Arts (NESTA), some of the European accelerators have adapted the business models of U.S. accelerators to account for differences in local funding and networking needs (Miller and Bound 2011). There are even government-funded accelerators. For instance, Start-Up Chile in Santiago (an accelerator found in just one of the six lists) attracts businesses from other countries and provides short-term funding to the startups but does not take equity stakes in the companies.

Table 3. Comparison of accelerators in the United States and abroad

	Organizational Structure	Access to Investors	Industry Focus
United States	Geographically based: Cohorts of small businesses are located in close geographic proximity to mentors and investors to reduce barriers to market entry.	Access to potential investors factors into location of accelerator.	Software design and mobile application development
Abroad	Event-based: Accelerator headquarters or branch offices host events around the region for small businesses. Investors are not necessarily in close proximity to the mentored businesses.	Access to potential investors is not always a factor in organization of accelerator events. Access is limited by distance between startups and investors.	Software design and mobile application development

Organizational structure and access to investors are some primary differences between accelerators in the United States and those located overseas (Miller and Bound 2011). For example, Startupbootcamp operates programs in alternating European cities. Some accelerators, like the London-based accelerator Seedcamp, are working to close the investment gap in startups between the United States and Europe. Seedcamp has worked with the British government to develop a visa that would allow accelerator founders from outside the European Union to relocate and work in the UK at a lower cost. Multistage investment funds have also been developed to help close this gap. Proponents view this investment strategy as a way to increase the "pipeline" of companies in need of investment from venture capitalists.

As noted by NESTA and NBIA, accelerators typically focus on startups pursuing software design and mobile applications. However, at least one of the top-cited accelerators listed in table 2 (DreamIT Ventures) specifically states on its website that its industry interests are broad-based and include anything that can be developed into a product in 3 months. Other accelerators in the list, such as Impact Engine, focus on software and mobile applications but in socially minded contexts, for instance civic ventures that help reduce the recidivism rate at jails or to promote multicultural meal-sharing. Basic questions, such as the most common industries that accelerators target, the rate of new program creation, and the geographic distribution of programs, are currently difficult to answer. A standardized definition is necessary in order to sort out these differences and provide a foundation for future research, as well as to answer deeper questions about this burgeoning startup assistance model, such as the average return on investment for accelerators.

Recent additions to the literature have begun to address the need for a widely shared definition. The articles indicate a convergence of thought around a defining set of characteristics. The report by NESTA, an innovation-focused research firm in the UK, provided the following list of accelerator characteristics (Miller and Bound 2011):

- An open but competitive application process.
- Provisions of pre-seed investment, typically in exchange for equity.
- A focus on small teams (generally including up to four members) rather than individual mentoring.
- Time-limited support that includes programmed events and intensive mentoring.
- Cohorts of startups rather than individual companies.

One of the most recognized sources of information on accelerators, —Seed-DB, —also advocates and applies most of these as criteria for compiling its database.[4] Collaborators on an inter-organizational effort called the Seed Accelerator Ranking Project largely concur with NESTA's approach (Cohen 2013; Cohen and Hochberg 2014). They provide a succinct definition of accelerators based on four characteristics:

> a *fixed-term*, *cohort-based* program, including *mentorship and educational components*, that culminates in a public pitch event or *demo day*.[5]

This definition also helps distinguish which aspects of accelerator and incubator programs are more typical of one or the other and which aspects are shared. The Venn diagram in figure 1 illustrates the unique and overlapping characteristics of the two kinds of organizations.

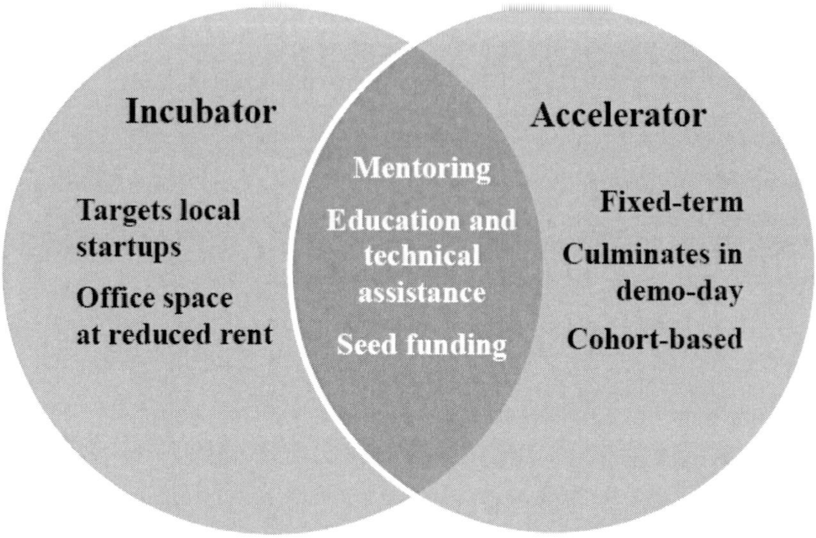

Figure 1. Venn diagram of incubator and accelerator characteristics.

Criticisms of Accelerators

The accelerator phenomenon is not without its critics. One important criticism is that their effectiveness at improving startups' rate of survival and success is unproven. Jared Konczal, policy analyst at the Kauffman Foundation, argues that some in the industry have misused statistical measures to present an overly positive assessment of early accelerator results (Konczal 2012). Among his criticisms are several statistical problems, including missing or inaccurate data, limited population and sample sizes of startups, and measures that may be heavily skewed by outliers (Y Combinator in particular is highlighted as an outlier). Other measurement and performance-evaluation problems that Konczal identifies include selection bias, introduced because accelerators select only high-potential startups, and inaccurate or incomplete reporting of costs and performance data by accelerators and their startups. Central to Konczal's criticism is the actual cost per job created; he contends that this cost is substantially higher than the amount reported by industry groups (e.g., Grasshopper Blog, Global Accelerator Network).

Although the statistical problems Konczal identifies are real, we argue later in this report that assessing the performance of privately funded for-profit accelerators using nonprofit and public criteria (i.e., job creation and equitable access to services) is simply not appropriate. The next section elaborates on this fundamental difference. But in summary, the difference between privately funded and nonprofit accelerators points to a more serious criticism of accelerators and one reason for this current study. There is considerable confusion among the many types of programs available for startups. In the absence of clear distinctions, entrepreneurs may not have sufficient information to make informed decisions when entering into agreements with accelerators. At the same time, policymakers may not have sufficient information to enact appropriate policies to, for example, promote or support the long-term viability of young entrepreneurs in the startup market.

Additional criticisms are leveled at specific subtypes of accelerators.[6] For instance, a number of large corporations, such as Microsoft, have founded their own accelerator programs. Some authors in the field are concerned about corporate accelerators due to previous experience with corporate incubation programs, which they argue may have created market bubbles because of corporations' fluctuating interest in investing (Bradford 2014; Konczal 2012; Trotter 2013). Other criticisms of corporate accelerators include mixed performance from the startup perspective. These span several important dimensions, including mentorship, appointment of accelerator directors, and accelerator incentive systems that reward short-term accelerator performance instead of long-term startup growth (Bradford 2014). Bradford echoes the problem of startups having insufficient information to make decisions; he identifies some accelerators' misrepresentation of deal terms as a problem that startups and corporate sponsors alike need to be aware of and address.

Toward an Accelerator Taxonomy

The literature to date has generally characterized accelerators by the services they provide and the programs through which they provide these services. Although this approach yields a useful starting point for research, there may be opportunities to define accelerators by more fundamental criteria. The definition cited above is based on accelerators as programs within an organization rather than as a business entity itself. Incubators and other startup assistance organizations that engage in the process of accelerating innovation may or may not call themselves accelerators, and many of the organizations that call themselves accelerators (as defined by Cohen and Hochberg) may be more aptly characterized as businesses in the seed-capital investment industry. Specifically, they provide innovation-acceleration services as a means of securing and enhancing their investments in a portfolio of startup companies. For instance, as will be discussed shortly, corporations and universities have both sponsored accelerators. Some of these programs meet the definition provided and most are absent from the inventories, but they are important variants of the accelerator phenomenon.

In addition, a definition based on services provided does not offer adequate foundation to determine whether and how public policy should respond. As private ventures motivated by profit, most accelerators are *not* primarily in the business of providing a public service to entrepreneurs, which is a critical difference between accelerators and other startup assistance programs that have their genesis as publicly funded programs. This profit objective is a defining characteristic of the early accelerators.[7] Although attention is often focused on the startups themselves and the more altruistic goal of helping them succeed, the for-profit accelerator business model is rooted in an investment strategy for developing a portfolio of seed-stage investments. The portfolio is designed to yield a predictable minimum return on its contents as a whole over a specific time frame, although this time frame may vary significantly across accelerators. Once a cohort of startups is selected, the accelerator enhances the prospects that its investment will be profitable by providing a rigorous program of training, mentoring, and technical assistance that gives the new startups the best chance for rapid and high growth.

Given the differences in accelerator-like organizations, a taxonomy of both key services and key foundational elements may help sort their benefits and their place within the innovation ecosystem. The methodological approach for developing a taxonomy of accelerators and similar startup assistance programs is summarized here and described further in appendix 1. The following steps were taken to develop the taxonomy:

1) Thoroughly review the literature to determine how the defining characteristics of accelerators relate to the accelerator business model.
2) Evaluate the differences between accelerators and incubators from the business plan perspective (a departure from prior classification efforts that evaluated accelerators only from the perspective of observed characteristics).
3) Reframe the observed characteristics within the business model framework.
4) Identify other organizations with accelerator services described in the literature and their distinct characteristics, including examples.

5) Create a classification chart that sorts organizations with accelerator services and characteristics into their respective organization type.

4. Classifying Accelerators

Overview and Goals of an Accelerator Taxonomy

Taxonomies are classifications of items into ordered categories based on shared characteristics, from general, observed characteristics to unique and potentially unobserved characteristics. They are frequently used in biology (e.g., to classify species), but the method can be applied broadly and will be useful for differentiating among startup assistance organizations. The first goal of the taxonomy is to help identify and organize fundamental and observable characteristics of accelerators and disaggregate them from characteristics that they share with similar types of organizations. Once an accelerator typology has been established, a second goal is to use this new taxonomy to classify existing accelerators, incubators, and similar organizations to help understand the marketplace for accelerator services.

What Makes Accelerators Unique: The Value Proposition and Business Model

Accelerators, incubators, and related entities may be differentiated by their value proposition and business model.[8] Previous characterizations have hinged on related attributes, such as accelerators' tendency to be for-profit, and the tendency of incubators and other startup assistance providers to be nonprofit organizations. This research argues that the business model and the value proposition are essential to the classification of an accelerator. The business model determines how the accelerator is structured, including its choice industry, and the value proposition determines what the accelerator offers to startups.

Value Proposition of an Accelerator

The value proposition describes the overall package of products, services, requirements, and costs associated with the seed capital provided to startups. The value proposition clarifies aspects of the business that create value for the customer, how customers experience that value, and what alternatives the customer has in the marketplace. The value proposition has five parts: (1) customer market, (2) activities, (3) rewards, (4) value experience, and (5) alternatives and differentiation. Each aspect is described below for accelerators.

1. Customer Market

Accelerators focus on a stage of new product development that occurs at the transition from invention to a commercial product. They serve three distinct markets concurrently during this stage:

- New and potential startup businesses comprising one or more entrepreneurs with rapid growth potential.
- Venture capitalists and other investors interested in funding startups once they emerge from the early stages of invention.
- Existing firms searching for new products or firms to acquire as part of their own business strategies.

2. Activities

Accelerators offer bundles of services to each of their customer markets. To the startups, they offer a bundle of services in exchange for equity stakes in the startups. These bundled services generally include brokerage services (e.g., demo days), mentoring and technical assistance services (e.g., human resource management), and a combination of cash and in-kind contributions (e.g., office space). Venture capital (VC) investors and established firms also receive brokerage services that keep them informed of viable investment and acquisition opportunities, respectively.

3. Rewards

An accelerator offers many potential rewards to startups.[9] These may include the acquisition of specialized knowledge (e.g., through the mentoring process), additional seed capital to proceed to the next stage of development, or even an initial public offering (IPO), and ongoing proof of their concept throughout the accelerator experience. The best accelerators and their founders have also developed strong brands, which tend to attract the best startups. This reputation for quality confers a distinct status on the startups that may benefit them in networking and searching for capital.

Rewards to VC investors and existing firms primarily include a reduction of real and opportunity costs associated with the search for new investment opportunities and the due diligence necessary to validate investment prospects.

This effectively reduces investment risk along two important dimensions: product risk and company risk.

4. Value Experience

Accelerators deliver a different value experience to each of their three markets.[10] However these value experiences are linked through the accelerators' brokerage function, which unites parties from each market. Thus, the value is an emergent and unpredictable reward based on the social capital created among the parties. Nonetheless, they are intentionally enhanced by such factors as the brand effect and cohort design of the accelerators.

5. Alternatives and Differentiation

Accelerators operate in a specialized niche with relatively few established alternatives[11] in terms of firms offering similar specialized services, rewards, and value experience for that particular industry.

- ***Differentiation:*** Accelerators may be differentiated along several dimensions, including the technology industries in which they specialize and their unique combination of brokerage, mentoring, and funding activities.
- ***Alternatives:*** Alternatives to participating in an accelerator include the "null" option, in which startups and investors navigate the transition from invention to product on their own without commercial or institutional assistance. The best-known institutional alternatives are business incubators, while other institutional alternatives generally provide a more limited subset of an accelerator's services. For example, Small Business Innovation Research/Small Business Technology Transfer (SBIR/STTR) Phase I grants provide capital for prototyping activities, while angel and venture investment "clubs" facilitate certain brokerage services that accelerators provide. But the systemic lack of coordination among the various partial programs and incubators' specific limitations may have helped create the niche market for accelerators.

Although any entity in the innovation ecosystem should seek to define its value proposition, the specific customer-market focus and benefits described here are proposed as unique foundations of an accelerator.

Accelerator Business Model

The business model describes how the accelerator is structured to obtain its goals, how it prices its products and services, and how it generates income and, in some cases, profit. Most of the existing accelerators to date have operated in the software or mobile applications arena, an industry characterized by relatively low capital requirements and short prototyping durations.

The accelerator's revenue assumptions are then built around rapid growth and large-scale markets. These factors drive a business model that allows the accelerator to operate sessions that last approximately 3 months with modest-size cohorts and relatively modest equity percentages (in the range of 5 to 8 percent) in the participant startups.[12]

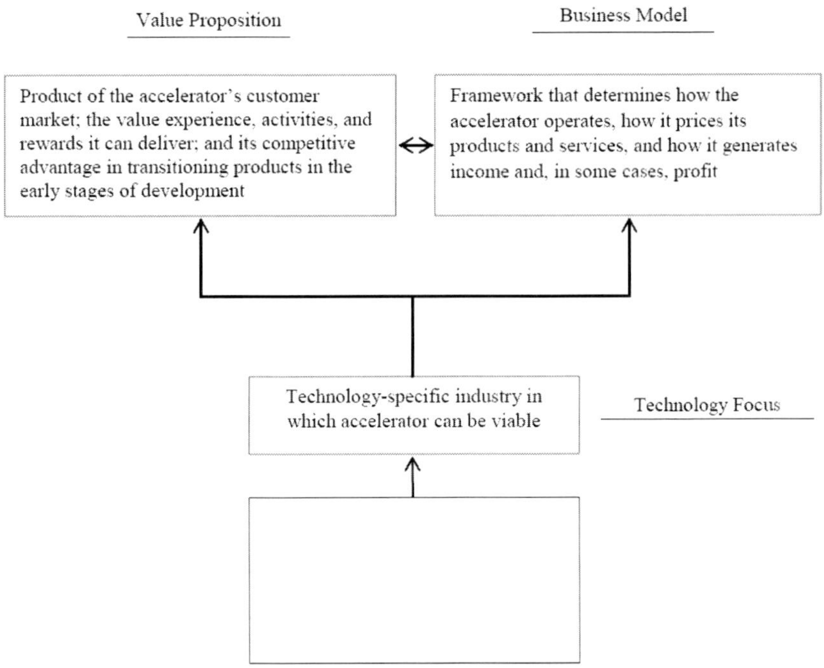

Figure 2. Foundational components of accelerators.

The focus on technology is a key factor that influences the accelerator business model, distinguishing accelerators from incubators and other startup assistance organizations. The term "technology focus" (or technology-specific industry) means that the accelerator is focused on startups in a relatively narrow range of related technologies. Figure 2 provides a conceptual map of how the business model and value proposition are influenced by the technology-specific industries in which accelerators work.

The accelerator's value proposition and business model must be reconciled to create value for customers and a profitable, sustainable operation for the accelerator's founder. Founders motivated by profit achieve this value through a technology focus in which they have past experience and current networks. The founder's knowledge and networks in a technology-specific industry are critical in evaluating accelerator applicants. The founder must determine whether the applicant's idea represents an innovation or potential innovation, whether or not that innovation has significant market potential, and the extent to which the startup team understands its specific industry. It is possible for the founder to hire mentors or take on partners with technical expertise and networks in other technology-specific industries. However, in the basic accelerator model, such hiring adds costs and increases risk, both of which the accelerator founder is likely to avoid. The founder and mentoring team must then evaluate the extent to which the accelerator can add value and accelerate a startup applicant's growth process through their mentors, technical assistance, and network connections.

If one looks at the original accelerators founded in the mid-2000s (many of which are shown in appendix 2) and assumes competitive market conditions, then the technology-specific industry of accelerators is determined by their founders' technology specialization, experience, and networks. If the founder has primary motives other than profit, then the business model can balance the expected profit with other motivations; this influences the extent to which the accelerator focuses on startups in a technology-specific industry. For example, many incubators list job creation among their end goals and measures of success, and some accelerators focus on civic ventures. Differences in founders' motivations appear to drive significant differences in the value proposition, which in turn drive different assumptions and characteristics of the business model, including the level of support offered and equity stake required, which further influence the value proposition. Through the process of building a business model and value proposition that achieve the founder's goals, variations in the functional characteristics of the accelerator emerge.

Taxonomy Content

Different types of organizations offer various subsets of services associated with accelerators and exhibit similar observable characteristics. For the purposes of this research, six primary types of organizations have been identified that offer accelerator-like services to startups, yet have significantly different business models due to the founder's primary motivation or the founding objectives of the organization. The six types of startup support organization are: (1) incubators, (2) venture development organizations (VDOs), (3) university accelerators, (4) proof-of-concept centers (POCCs), (5) corporate accelerators, and (6) the primary focus of this paper, innovation accelerators. They are each described briefly.

1. Incubators

According to NBIA, business incubation is a business-support process that helps launch startup and fledgling companies by providing entrepreneurs with an array of needed resources and services.[13] These services are usually developed or orchestrated by incubator management and offered in the business incubator and through its network of contacts. A business incubator's main goal is to produce successful firms that will be financially viable and free-standing when they leave the program. These incubator graduates have the potential to create jobs, revitalize neighborhoods, commercialize new technologies, and strengthen regional economies. Critical to the definition of an incubator is the provision of management guidance, technical assistance, and consultation tailored to young, growing companies. Incubators usually also give clients access to appropriate rental space and flexible leases, shared basic business services and equipment, technology support services, and assistance in obtaining the financing necessary for company growth.

2. Venture Development Organizations

According to the Regional Innovation Acceleration Network (RIAN), one of the more recent and most successful technology-based economic development interventions to emerge for regional innovation is an adequately resourced regional venture development organization.[14] A VDO is a public or nonprofit organization that contributes to economic development by providing a portfolio of services, including

- assisting in the creation of high-growth companies;
- providing expert business assistance to those companies;
- facilitating or making direct financial investments in companies; and
- accelerating the commercialization of technology.

A high-performing VDO draws on the existing strengths of the region's innovation system and develops programs and initiatives targeted to overcome the system's weaknesses. Properly structured VDOs have the ability to work with a wide cross-section of the key assets of their particular regional innovation systems and the flexibility to adapt their portfolios of services to meet the specific needs of individual commercialization opportunities or ventures. Organizations identified as VDOs by RIAN include such groups as JumpStart in northeast Ohio and i2E in Oklahoma.

3. University Accelerators

University accelerators are educational nonprofits that accelerate the development of student entrepreneurs and innovation at universities throughout the United States. University accelerators typically provide seed grants to support students through the early stages of development. Unlike for-profit accelerators, university accelerators do not take equity stakes in student-founded companies, and they are typically agnostic when it comes to technology focus. Some university accelerators, such as StartX at Stanford, extend services to faculty and alumni, as well. University accelerators provide the same range of services as other accelerators, including mentoring, technical assistance, use of facilities, and networking, usually including a demo day.

4. Proof-of-Concept Centers

Proof-of-concept centers accelerate the commercialization of innovations developed by university faculty and staff, and help move these innovations into the marketplace. POCCs provide seed funding for novel, early-stage research that most likely would not be funded by other conventional sources. Two examples of POCCs are iGreen New England Partnership and the Igniting Innovation (I2) Cleantech Acceleration Network in Orlando. POCCs facilitate and foster the exchange of ideas between university innovators and industries via the various mentors associated with the center. From a business-plan perspective, a POCC represents an investment by a university in improved technology transfer. In terms of the value proposition, POCCs provide a "collection of services to improve the dissemination and commercialization of new knowledge from universities in order to spur economic development and job growth" (Bradley, Hayter, and Link, p.3, 2013).

5. Corporate Accelerators

These accelerators engage in the provision of seed capital and various combinations of mentoring, technical assistance, networking, and facilities to entrepreneurs, inventors, and startup teams to advance certain goals of the corporate or institutional parent. Corporate accelerators (examples are shown in table 4) grow and manage portfolios of complementary startups to accelerate innovation and gain a competitive advantage. CorpVenturing assists Global 5000 companies with investing in innovation strategies and it suggests the following objectives as inherent to a corporate accelerator:

- "Find next-generation products in your industry that you can help commercialize.
- Create an ecosystem of users and customers for your key products.
- Drive innovation at a much faster pace than is possible internally.
- Create growth options by taking stakes in interesting companies.
- Gain a window into the technologies and business models that will be winners.
- Profitably leverage your existing scale, distribution, and relationships into additional value."[15]

Table 4. Examples of corporate accelerators

Accelerator	Corporate Sponsor
Siemens Technology Accelerator	Siemens
Nike+ Accelerator	Nike, Inc.
Citrix Startup Accelerator	Citrix
Media Camp Academy	Turner Broadcasting System and Warner Bros. Entertainment, Inc.
The Microsoft Accelerator	Microsoft, powered by Techstars
Volkswagen Electronics Research Lab Technology Accelerator	Volkswagen Group of America
Kaplan EdTech Accelerator	Kaplan, Inc., powered by Techstars

Because corporate accelerators are driven by different motivations, their business models differ from those of other accelerators. However, they generally work with technologies at the same stage of development and offer similar services to startups.

6. Innovation Accelerators (Private, for-Profit)

Innovation accelerators are stand-alone, for-profit ventures in the business of:

- Identifying cohorts of promising startup companies with rapid, high-growth potential.
- Making seed-stage investments in those companies in exchange for equity.
- Engaging in innovation-acceleration activities with these companies to help them obtain next-stage funding.
- Cashing out for a profit when these companies are acquired or have successful IPOs.

The business model of innovation accelerators is clear. Cohort size drives a probability model for each cohort and for the innovation accelerator's startup portfolio. This in turn allows the accelerator to balance the cohort size with anticipated return on investment across each cohort and the portfolio as a whole.

Taxonomy Structure

The six types of organizations offer a variety of services to startup companies looking for assistance. These organizations differ in key ways—in terms of program structure. operational features, financial capital and intellectual property (IP) rights and rules, and social capital and network-building opportunities. An accelerator founder's selection among them depends on the business model and value proposition of the organization. For simplicity's sake table 5 groups the characteristics into just two types of organizations — whether the organization's services and characteristics are more like an accelerator or incubator, as defined in the literature review.

The six organization types also display a range of primary motivations for promoting the success of startup companies. Table 6 organizes the entities across a set four motives, some private and some public. An organization founded with the objective to commercialize faculty and student research or generate local economic development are examples of "public good" motivations, while an organization that intends to further a corporate innovation strategy or generate profit from investment in startups are examples of a "private good" motivation.

Table 5. General characteristics of organizations that engage in startup assistance

	Characteristic	Incubator-like organization	Accelerator-like organization
Program Structure	Startup Selection Process	Competitive – based on available space and resources	Competitive – essential to business model
	Technology Focus	Broad	Narrow
	Stage of Technology	Seed and broad range of stages	Seed and pre-seed
	Cohort Structure	No	Yes—essential to business model
	Program Duration	More than12 months; average of 33 months	Less than 12 months; average of 3 months
Operational Features	Office / Lab / Flex Space & Equipment	Yes	Varies
	Mentoring	Yes	Yes
	Technical Assistance	Yes	Yes
	Post-Program Support	Varies	Varies
Financial Capital & IP Rights	Seed Funding	Varies – not typical	Yes—essential to business model
	Equity Stake	No	Yes—essential to business model
	Intellectual Property (IP) Conditions	Varies by organization's IP policies	Varies by equity agreement

	Characteristic	Incubator-like organization	Accelerator-like organization
Social Capital & Networks	Networking with Angel / VC & Corporate Investors	Varies according to staffing and resources	Yes—important to business model and value proposition
	Networking with Potential Customers & Suppliers	No	Yes—important to business model and value proposition
	Demo Day	No	Yes
	On-site Participation Required	Varies	Varies
	Cohort & Alumni Networking	Varies	Yes

The six organization types are associated with one or more objectives in the table. Identifying a primary motive is not to suggest that university accelerators, for example, are more interested in promoting university research than generating profitable startups. All organizations providing startup assistance likely want to see their participants succeed. However, the primary motive of a university starting an accelerator is one of commercializing research and generating economic development, rather than making a profit from its student startups.

Table 6. Unique characteristic of organizations that engage in startup assistance

<table>
<tr><th rowspan="2">Broad Category</th><th rowspan="2">Objective</th><th colspan="2">Organization Types</th></tr>
<tr><th>Single Primary Objective</th><th>Multiple Primary Objectives</th></tr>
<tr><td rowspan="2">Nonprofit/ Public Good</td><td rowspan="2">Commercialization of university and sponsored research
Local/regional economic development</td><td rowspan="2">Proof-of-Concept Centers
Incubators</td><td>University Accelerators</td></tr>
<tr><td rowspan="2">Venture Development Organizations Social Accelerators</td></tr>
<tr><td>For-Profit/ Private Interests</td><td>Profit from investment in startups
Further corporate innovation strategy</td><td>Innovation Accelerators
Corporate Accelerators</td></tr>
</table>

A seventh type of business assistance organization, *social accelerators,* also appears in table 6. They are quite rare and display a mix of founder motivations that bridge public and private goods. These experimental accelerators may seek profit while relaxing aspects of the business model to accommodate objectives that advance the public good. They may also be founded for the purpose of accelerating nonprofit and social enterprise startups while adopting features that promote accelerator profit. The ARK Challenge in Arkansas is one potential example of a social accelerator. It is focused on Web-based and mobile technologies in the financial information, health information, and government services sectors. It takes equity stakes in its startups, but is also organized as a nonprofit, has received federal funding, and seeks in part to promote job creation in the region.

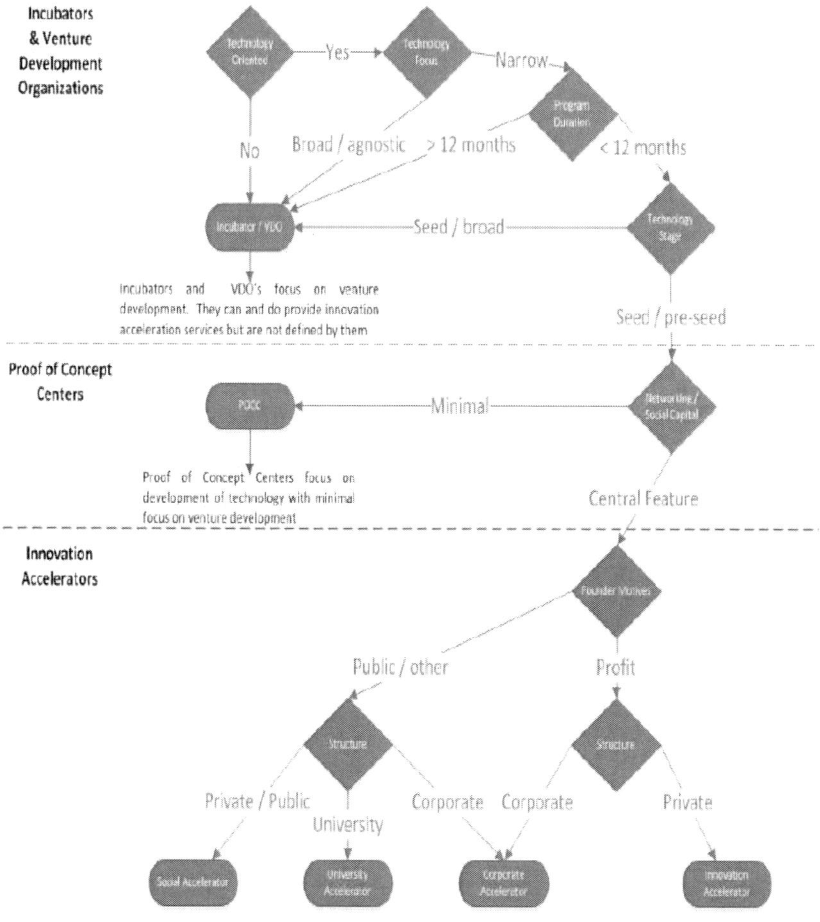

Figure 3. Taxonomy of organizations that engage in startup assistance.

On the one hand, two accelerators may present different general and observed characteristics— for example, one provides office space and the other does not—but they are still innovation accelerators because they share the same primary objective and the same basic business model that maximizes profit from investment in startups. On the other hand, an incubator and an innovation accelerator may both provide office space, but remain different types of organizations, because their organizational objectives are different and they have substantially different business models. To classify startup assistance organizations requires identifying the underlying differences in the founder's motivation and business model, which in turn results in differences in the observed characteristics.

Pulling together the foundational aspects, such as technology focus and key observed characteristics from the accelerator definition presented in the literature review, the taxonomy shown in figure 3 is a first attempt at accelerator classification. It can be used to facilitate classification of a specific organization. Walking through each decision point (green), a user can best estimate whether the organization is an incubator or VDO versus a POCC. Once in the arena of accelerators, the next series of decision points (shown in orange) lead to classification among the current variants of accelerators. The differentiation among the three levels is marked with dotted lines to represent the flexibility in the definitions of these organizations.

Innovation Accelerators

According to the definition of Cohen and Hochberg (2014), an accelerator program is a *fixed-term*, *cohort-based* program, including *mentorship and educational components*, that culminates in a public pitch event, or *demo day*. The present research proposes a modification to this definition which distinguishes between innovation acceleration as a *program* and an innovation accelerator as a *business model*. In this latter sense, innovation accelerators are defined as "business entities that make seed-stage investments in promising companies in exchange for equity as part of a *fixed-term*, *cohort-based* program, including *mentorship and educational components*, that culminates in a public pitch event, or *demo day*." While the immediate goal of the accelerator is to help their startup companies obtain next-stage funding, their primary longterm goal is to make a substantial profit when those companies are acquired or have successful IPOs.

Cohen (2013), Cohen & Hochberg (2014), and this study identify similar defining characteristics for accelerators. The question remains as to whether or not the characteristics of the programs are defining features of the business entity. Cohen (2013) argues that the limited duration of accelerator programs is perhaps their single most important defining feature, an assumption that this study has also reflected. However, from the perspective of a business model and the value proposition offered to startups, the limited duration is not driven solely or even primarily by the accelerator founder's altruistic concern for the well-being of the startup. Such concerns may be genuine and strongly felt by a founder, but the limited duration is also a function of the business model economics. In addition to the "tough love" benefits that a short, intense, focused time frame offers startups, the limited duration is essential to controlling costs and increasing the number of startups in the accelerator's portfolio. This in turn increases the expected profit of the accelerator by increasing the probability of one or more high value exits into the marketplace. Insofar as policymakers and startups look to definitions proposed by researchers to make decisions about if and how to engage with accelerators, an understanding of both their programmatic and organizational characteristics are important.

5. Data and Metrics

What Should be Measured: Accelerator Performance and Evaluation

Different motivations at the founding of an accelerator, incubator, or other business development organization affect how success is defined. For example, an incubator could be evaluated for the number of new businesses that start and remain local, whereas an accelerator could be evaluated for the speed at which a new business is acquired. The metrics discussed in this section are applicable to business performance generally. To measure the effectiveness of accelerators, two questions about metrics need to be answered:

1) What metrics will help determine whether accelerators are successful?
2) What metrics will help determine how accelerator success varies by type of accelerator or business development organization?

Two reports in the literature review are focused on measuring accelerator performance, both of which were published in 2013. The metrics identified in table 7 are based on the discussion in the literature about the merits of evaluating performance of Canadian accelerators (Baird, Bowles, and Lall 2013; Caley and Helen 2013). In the short-term, metrics such as the startup's status after the program (e.g., operating or closed) are important for measuring performance. More specifically, Baird et al. (2013) suggest that the startup must raise $500,000 or more in investments to be considered a success. Over the longer term, performance measures should include the startup's growth, with metrics such as increased sales and investor rate of return. The articles also discuss the performance of the accelerator. In the short term, accelerators can be measured on the acceptance rate and frequency with which graduate startups are acquired. In the long run, Baird et al. (2013) suggest accelerators evaluate their internal rate of return and sources of funding—particularly if the accelerator does not take equity stakes in its startups.

Table 7. Short- and long-term metrics for accelerators and their startup firms

Time Horizon	Accelerator Metrics	Startup Metrics
Short-Term (program duration plus 6 months)	Number of applicants Number of participants (cohort size) Number of investors at demo day Percentage receiving next-stage funding Percentage acquired Percentage failed	Operational status (operating, closed, acquired) Number of financial investments or number of investors Size of financial investments Number of customers gained
Long-Term (expected cash-out in 3–7 years)	Sources of funding (series or portfolio) Performance distribution (cohort or portfolio) Internal rate of return (cohort or portfolio) Network metrics (partnerships, etc.)	Sales or revenue Number of employees Rate of return to investors Stock prices (if applicable)

One shortcoming of these and other metrics- and performance-based evaluations is that they are not able to account for the scenario of a "successful failure." A startup may not last more than a few months; however, if this quickly demonstrates that its idea was not viable, it conserves resources for other ventures. This phenomenon is not captured in the current metrics, but the accelerator and the startup may consider such a scenario to be a valuable outcome. From the accelerator's perspective, it helped the startup and investors avoid going down an unprofitable path, and the accelerator could devote resources to helping the startup redesign the venture or develop a new venture that would be more successful. From the startup's perspective, the accelerator helped identify areas of weakness that could have resulted in a failed enterprise with potentially severe financial consequences.

The metrics in table 7 represent a good start for evaluating whether or not a specific accelerator is successful and whether or not accelerators in general are successful in their goals. A significant amount of overlap exists between the metrics that have been proposed for evaluating incubators and those proposed for accelerators (Voisey et al. 2006). To evaluate accelerator performance, especially as it compares to other similar entities, the evaluation design needs to consider program characteristics. Some characteristics that may be related to success, or the scale of success, in accelerator programs include:

- Intensive format of mentoring and business skills training.
- Program length.
- Historical connections to investors.
- Team-based startups rather than individuals.
- Technology-focused industry.

The results of an evaluation that takes into account the program design could help determine the value of accelerators as a type of "intervention" in small business development. Moreover, an evaluation that assesses long-term startup metrics will help determine accelerators' levels of success and their economic impact. Similar to determining the success of a single accelerator, determining the success of accelerators as a group or a class of business development organization requires several years before the metrics can provide useful indicators of long-term success.

What Can Be Measured: Data Landscape and Limitations

Currently, little public data is available to evaluate accelerators and their startups, in part due to the industry's propriety nature. The data on startup survival and success that are compiled typically lack consistency and an external verification process. Government and business data sources may provide the consistency and validity needed, but accelerator participants are not readily identifiable. Similarly, there is no reliable, centralized source of data on the accelerator programs themselves (e.g., number of graduates and number of startups that acquire a subsequent round of funding after the program), though entrepreneur and accelerator industry groups provide some centralized information. The table below provides an overview of the currently available types of data sources and their contributions to evaluating accelerators. Additionally, the Seed Accelerator Ranking Project (discussed in the literature section) may represent the next stage in the development of sound, reliable data for analysis and policy. This project seeks to move beyond simple lists to develop rigorous quantitative and qualitative metrics around the characteristics, activities, and performance of accelerators (Cohen and Hochberg, 2014). More attention needs to be placed on collecting data on the performance and economic impact of accelerators to adequately assess whether they are more, less, or equally effective as other business development organizations. If data collection cannot be organized around a shared definition (like the one proposed here), it should be done under the guidelines of clear methodologies so that different sources can be reconciled.

Table 8. Overview of existing types of data sources and their benefits

Data Source	Description
Open Source	These sources provide free and easy access to data most directly associated with accelerators, but they are typically unverified by an external reviewer.
Techstars	The Techstars statistics database provides a complete listing of companies it has funded and other, more detailed information. The variables provided include current status (active, acquired, or failed), number of employees in each company, and average amount of funding received. It also reports this information by cohort in a summary table and cohort-specific tables that show how many companies were in each cohort, and the average amount of funding acquired by the cohort overall.
Seed-DB	Seed-DB provides statistics and listings of companies in a large number of accelerators. Seed-DB acknowledges that its database is incomplete and may contain outdated information, as much of the data are pulled from other sources such as CrunchBase. The statistics available are similar to those provided by Techstars.
CrunchBase	CrunchBase is a self-reported database of companies. Many accelerators and their startup graduates have profiles on CrunchBase that provide information on the founders, date of founding, current activity, and (occasionally) funding information. Because it is self-reported, it is not possible to guarantee that the information is complete or up to date. Daily e-mails or feeds provide news and information about recent activities. Downloadable spreadsheets are updated monthly.
Proprietary Market- Based Data Sources	Market-based data sources provide economic and financial information on startups after they have left the accelerator. The data are often proprietary, so the sources sometimes charge fees for the information.
Dow Jones VentureSource Database	The database provides accurate and comprehensive data on venture-backed companies. It includes investors and executives in every region, industry, and stage of development around the world. VentureSource is designed to help venture capitalists, corporate development executives, investment bankers, and service providers.

Data Source	Description
New York Stock Exchange	The exchange provides stock and financial information on startups that have gone public. It also may provide some information on investors associated with the startups.
Dun and Bradstreet (D&B)	This company provides information on businesses that have applied for business credit and have a DUNS number. It also provides basic information about these businesses along with information on their annual sales.
National Establishment Time Series (NETS)	This database offers time-series data on business establishments, including industry NAICS codes, sales, and other information extracted from D&B, plus information on parent firms, acquisitions, mergers, and company relocations.
Government	Government sources of data track a variety of important innovation indicators that may be manually linked to information on accelerator graduates.
Patent Filings	Patent filings show whether a startup has progressed on its goal or has expanded.
SBIR Awardees	SBIR awards provide valuable information on the success of a startup because they show that startups are active in the market and competitive for funding. Because many SBIRs also require annual reporting, it might be possible to obtain more specific information regarding the activities of startups with grants from the funding agency.
Quarterly Census of Employment and Wages (restricted)	This program provides information on employment and payroll on a quarterly basis. The data are restricted to qualified and authorized researchers. When reporting results using these data, care must be taken to avoid violating Privacy Act provisions.

6. Summary and Policy Implications

Accelerators are part of an innovation ecosystem that is the focus of active research. This study has reviewed the available literature and data on accelerators to describe the accelerator phenomenon, distinguish accelerators from other entities providing similar services, and examine potential metrics and data sources that might be used to measure and monitor the activities of accelerators. The research confirms that the study of accelerators is still nascent and often produced as a by-product of research in related fields, such as innovation and entrepreneurship, where the amount of research is more robust. As this report and prior research have shown, distinctions between organizations often blur as startup assistance is designed to fit local conditions and meet local opportunities. Incubators, venture development organizations, corporate accelerators, social accelerators, and other variants surveyed in this study offer similar services to startups. Given the fluidity in services, a key distinction made here is between accelerator programs, which may be hosted by a range of entities, and innovation accelerators, which are a specific type of business entity with a clear value proposition for customers and business model designed to generate profit. Foundational elements that distinguish innovation accelerators from other entities that provide accelerator-like services include:

- the founder's motivation or objective in starting the organization, because different profit and public motivations shape the business model developed and the services offered; and
- the extent to which the organization is focused on specific technologies and development stages, which allows the organization to provide specialized technical assistance that increases its value proposition to startups.

Cohen and Hochberg (2014) provide a useful definition of accelerator programs which focuses on their operational characteristics. This report used that definition and adds to it in order to more narrowly define innovation accelerators as *business entities that make seed-stage investments in promising companies in exchange for equity* as part of a fixed-term, cohort-based program, including mentorship and educational components, that culminates in a public pitch event or demo day. The definition is based on structural characteristics that influence the business model and value proposition of accelerators, rather than operational characteristics that are determined by the business model and value proposition. Such a distinction helps suggest specific policy implications from this research.

Be Mindful of Accelerator Subtypes when Measuring Performance

As this report brings to light, there are inherent problems associated with measuring accelerator performance solely with a set of observed characteristics, or using outcome metrics such as job creation that are associated with nonprofits or government sponsored programs. The main problem is that these metrics often do not measure what is important to the accelerators, their startups, or subsequent investors, especially in situations where accelerator funding is entirely private. Indeed, these programs may be more interested in their startup exclusion criteria (i.e., high selectivity) and the speed with which startups can be abandoned or invested in more heavily. As a result, some metrics may be applicable to startup assistance programs generally, but policymakers and researchers should be mindful of the accelerator subtype in question and moderate expectation accordingly. Specifically, accelerator metrics should address participation, process and performance for four distinct constituencies: startups, accelerator programs, accelerator sponsors, and follow-on investors. Such metrics would allow for true comparisons across the various types, business models, industries and so forth. They may also help distinguish new accelerator models, as the market for accelerators continues to grow.

Data Are Needed to Report Relevant Metrics for Accelerators

As the data landscape survey suggests, the available data upon which new metrics might be developed is limited. Open source data are not validated and validated data are often restricted for use through regulations or cost. Moreover, important attributes of accelerators, such as their investor networks and ability to turn social capital into vital business assets, are particularly amenable to social network analysis metrics. Such metrics may require the ability to easily connect different data sources at the organizational level, or even individual level. While sources such as CrunchBase and Seed-DB seek to fill a data gap with open source data models, they do not do so for the purposes of informing public policy or academic research. The Seed Accelerator Ranking Project moves in the direction of academic research. However, it is currently unclear whether or not those data would be available for public research, under what conditions, and at what cost. Specific data needs will arise out of the development of a clear set of metrics. Government administrators may work in partnership with current data collection efforts to further the reliable use of these data for academic and policy research on accelerators and related topics.

Areas of Further Research Should Foster Public Objectives

Innovation accelerators have emerged from the private sector as viable companies for turning a profit. Accelerator programs have now also developed from academia and local nonprofit organizations. As such, accelerators represent a broad-based market response to the real and perceived need and opportunity associated with providing early stage startups with a well-defined set of services and network opportunities. They reduce the costs an entrepreneur faces–in time and resources–during the commercialization process. Nonetheless, many factors affect the viability of a startup, which causes innovation accelerators (if not other accelerators) to pick and choose those with the best chance of high-growth success. There is still much that remains unknown about accelerators in terms of potential market failures in startup assistance, which some variants are positioned to address. For example, much could still be learned about who participates in accelerators and why; whether the geographic distribution of accelerators provides adequate access to potential participants; and how the acceleration process differs across industries, particularly in national priority industries like advanced manufacturing. Rigorous empirical research into this emergent phenomenon is scant, and several topics could provide valuable information to help local and federal governments determine their role in innovation acceleration.

Targeted Pilot Projects Could Proceed

While there is a need for more research, this obstacle is not one that needs to prevent governments from engaging in targeted pilot projects that support innovation acceleration in key industries, for example, advanced manufacturing. Even with the limited available data, the business model makes it clear that accelerator will face greater challenges in industries where the time to prototype a product is longer, the process is more complex, and the capital requirements are higher. Well-designed pilot projects could help determine appropriate levels of support and mechanisms for delivering such support equitably to accelerator programs across the range of business models. Such policies will require evidence of what works outside of the early business models of innovation accelerators. Conversely, government support of a limited expansion of participants (and data collection) in a well-functioning innovation accelerator may provide another useful pilot project. Targeted accelerator pilot projects could develop evidence while researchers continue to develop new datasets, metrics, and empirical research on the broader impacts of accelerators.

REFERENCES

Adkins, Dinah. (2011, June). "What are the New Seed or Venture Accelerators?" *NBIA Review*. Retrieved June 5, 2014 from http://www.nbia.org/resource_library/review_archive/0611_01.php.

Adkins, Dinah. (2014). "Business Accelerators and Business Incubators, Part 1." White paper. Badir Program for Technology Incubators, Saudi Arabia. Retrieved August 11, 2014 from http://www.badir.com.sa/en/resources /downloads.

Allen, David, and Syedur Rahman. (1985). "Small Business Incubators: A Positive Environment for Entrepreneurship." *Journal of Small Business Management,* 23, 12-22.

Auerswald, Philip, and Lewis M. Branscomb. (2003). "Valleys of Death and Darwinian Seas: Financing the Invention to Innovation Transition in the United States." *The Journal of Technology Transfer*, 28.3-4, 227-239.

Baird, Ross, Lily Bowles, and Suaraph Lall. (2013, June). "Bridging the 'Pioneer Gap': The Role of Accelerators in Launching High-Impact Enterprises." Aspen Institute. Retrieved June 5, 2014 from http://www.aspeninstitute.org/publications/bridging-pioneer-gap-role-accelerators-launching-high-impact-enterprises.

Barnes, Cindy, Helen Blake, and David Pinder. (2009). Creating & Delivering Your Value Proposition: Managing Customer Experience for Profit. London: Kogan Page Publishers.

Bøllingtoft, Anne, and John Ulhøi. (2005). "The Networked Business Incubator–Leveraging Entrepreneurial Agency?" *Journal of Business Venturing,* 20, 265-290.

Bradford, Jon. (2014, March). Corporate-Run Startup Accelerators: The Good, the Bad, and the Plain Ugly. *tech.* Retrieved June 5, 2014 from http://tech.eu/features/779/corporate-runstartup-accelerators-good-bad-plain-ugly/.

Bradley, Samantha R., Christopher Hayter, and Albert Link. (2013). "Proof of Concept Centers in the United States: An Exploratory Look." Working Paper 134, Department of Economics Working Paper Series, University of North Carolina, Greensboro.

Caley, Elizabeth and Kula Helen. (2013, July). "Seeding Success: Canada's Startup Accelerators." MaRS, Toronto, Canada. Retrieved August 11, 2014 from http://www.marsdd.com/app/uploads/2013/07/Seeding-Success_v94.pdf.

Cárdenas, Fernando. (2012, June). *Business Elevators: An Innovative Model for Accelerating Growth of SMEs in Developing Market.* Master's degree thesis, Massachusetts Institute of Technology, Cambridge, MA. Retrieved June 5, 2014 from http://hdl.handle.net/1721.1/72937.

Christiansen, Jed. (2009). *Copying Y Combinator: A Framework for Developing Seed Accelerator Programmes.* MBA dissertation, University of Cambridge, United Kingdom.

Retrieved June 5, 2014 from http://www.seed-db.com/about/view?page= research.

Clark, Jennifer. (2009). "Coordinating a Conscious Geography: The Role of Research Centers in Multi-scalar Innovation Policy and Economic Development in the US and Canada." *The Journal of Technology Transfer,* 35.5, 460-474.

Cohen, Susan. (2013)."What Do Accelerators Do? Insights from Incubators and Angels." *Innovations*,8.3-4, 19-25.

Cohen, Susan and Yael Hochberg. (2014). "Accelerating Startups: The Seed Accelerator

Phenomenon." *Social Science Research Network.* Retrieved August 7, 2014 from http://ssrn.com/abstract=2418000.

CrunchBase (2007). Startup Search Page. Retrieved August, 2014 from www.crunchbase.com.

Fishback, Bo, Christine Gulbranson, Robert Litan, Lesa Mitchell, and Marisa Porzig. (2007). "Finding Business 'Idols': A New Model to Accelerate Startups." Ewing Marion Kauffman Foundation, Kansas City, MO.

Hoffman, David L., and Nina Radojevich-Kelley. (2012). "Analysis of Accelerator Companies: An Exploratory Case Study of Their Programs, Processes, and Early Results." *Small Business Institute Journal,* 8.2, 54-70.

Isabelle, Diane. (2013, February). "Key Factors Affecting a Technology Entrepreneur's Choice of Incubator or Accelerator." *Technology Innovation Management Review.* Retrieved August 11, 2014 from http://timreview.ca/article/656.

Jackson, Deborah. (2011). "What is an Innovation Ecosystem?" Engineering Research Centers, National Science Foundation, Arlington, VA. Retrieved June 5, 2014 from http://ercassoc.org/content/what-innovation-ecosystem

Konczal, Jared. "Evaluating the Effects of Accelerators? Not So Fast." *Forbes.* August 8, 2012. Retrieved June 5, 2014 from http://www.forbes.com/sites /kauffman/2012/08/08/evaluating-the-effects-of-acceleratorsnot-so-fast/.

Maia, Catarina, and João Claro. (2013). "The Role of a Proof of Concept Center in a University Ecosystem: An Exploratory Study." *Journal of Technology Transfer,* 38.5, 641-650.

Markman, Gideon, Phillip Phan, David Balkin, and Peter Gianiodis. (2005). "Entrepreneurship and University-based Technology Transfer." *Journal of Business Venturing,* 20, 241-263.

Miemis, Venessa. (2013, February 28). 62 Top Startup Accelerators in the US, Canada & Beyond [blog post]. Emergent by Design. Retrieved February 4, 2014 from http://emergentbydesign.com/2013/02/28/62-top-startup-accelerators-in-the-us-canadabeyond/.

Miller, Paul, and Kristen Bound. (2011). *The Startup Factories: the Rise of Accelerator Programmes to Support New Technology Ventures.* Discussion Paper. National Endowment for Science Technology and the Arts (NESTA), London, UK.

Phan, Phillip, Donald Siegel, and Mike Wright. (2005). "Science Parks and Incubators:

Observations, Synthesis and Future Research." *Journal of Business Venturing,* 20, 165- 182.

Seed-DB. (2009). Seed Accelerators. Retrieved February 4, 2014 from http://www.seeddb.com/accelerators.

Silva, Rick, David Allen, and Richard Traystman. (2009). "Maturing Early-Stage Biomedical

Research: Proof of Concept Program Objectives, Decision Making and Preliminary Performance at the University of Colorado." *Medical Innovation & Business,* 1.1, 52-66.

Sparks, Erin. (2013, August). "Top Trends in State Economic Development." National Governors Association. Retrieved from http://www.nga.org/files /live/sites/NGA/files/pdf/2013/1308TopTrendsinStateEconDevP aper.pdf .

Tech Cocktail. (2012, August). "A Guide to Choosing the Best Accelerator for Your Tech Startup." *Tech Cocktail.* Retrieved February 4, 2014 from http://tech.co/reports/startupaccelerator-report-2012.

Trotter, Alasdair. (2013, August) "Six Lessons for Corporations Building Innovation Accelerators." *Strategy & Innovation.* 11.2. Retrieved June 5, 2014 from http://www.innosight.com/innovation-resources/

Voisey Pam, Lynne Gornall, Paul Jones, and Thomas Brychan. (2006). "The Measurement of Success in a Business Incubation Project." *Journal of Small Business and Enterprise Development,* 13.3, 454 – 468.

Webbmedia Group. Big List of Incubators, Accelerators. *Webbmedia Group*. Retrieved February 4, 2014 from http://webbmediagroup.com/list-of-incubators-and-accelerators.

APPENDIX 1: METHODOLOGY AND RESEARCH PROTOCOLS

Literature Review

During fall 2013, Optimal conducted an extensive Internet search of published and unpublished research using variations and combinations of the following key words: “innovation,” “accelerator,” “incubator,” “venture capitalists,” “innovation ecosystem,” and “entrepreneurship.” The search started with peer-reviewed journal articles, was then expanded to working papers and government, nonprofit and business reports, and was finally extended to include gray literature, such as news articles, blogs, and the websites of specific accelerators and incubators. The literature was then categorized by topic (e.g., policy-related topic, core accelerator research, incubator definition articles, proof-of-concept center information). It was reviewed for trends in definitions across accelerators and incubators, characterizations of innovation and the innovation ecosystem, and trends in key characteristics of accelerators, their founders, and their purposes. Additional key research articles related to the definition of an accelerator were added to the review in the summer of 2014 due to its relevance for evaluating trends in a nascent research field.

Compendium of Accelerator Inventories

Several organizations and media groups have published inventories of accelerators in the form of rankings, working lists, websites for centralized advertising to startups, and more. Optimal selected six inventories to generate the list of accelerators in appendix 2:

- Seed-DB’s online database of accelerators;
- The top 62 accelerators by country, produced by Emergent By Design blogger Venessa Miemis;

- A guide to choosing the best accelerators among 70, produced by Tech Cocktail;
- A list compiled by Webbmedia Group including accelerator and incubator programs that have invested in startups in the past 12 months;
- A working list of U.S. and non-U.S. accelerators compiled by NESTA; and
- A short list of popular accelerators compiled by the founder of Seed-DB, Jed Christiansen, which was replaced by the full list on Seed-DB during this project.

Citations for these data sources are found in this report's references. The inventories represent some continually updated lists, which Optimal accessed during winter 2014, and some static lists published between 2011 and 2013. The compendium in appendix 2 represents all accelerators that were identified in at least two of the six inventories. This process was meant to reduce the number of defunct accelerators and the number of incubators and other entities that self-identify as an accelerator but have less of a consensus around that identification.

The six inventories include different data points about accelerators, such as year founded, founder(s), city, state, country, focus, website(s), or the funding provided to startups. The compendium contains the two most commonly available data points: year founded and location. Any discrepancy in location was resolvable through the inventories because most of them contained the location. In the event that the year founded could not be verified in at least two inventories, or the inventories produced an inconclusive result, Optimal consulted two additional data sources during spring 2014. First, the online startup database CrunchBase was used to provide a second verification source or to resolve inconclusive results from cross-referencing within the inventories. If the discrepancy or missing data could not be resolved with CrunchBase, Optimal visited the accelerator's website or LinkedIn webpage. Where a founding date was identified on an accelerator webpage, Optimal did not require an additional data source for verification, however very few of the remaining dates that were undetermined could be found on the websites. All remaining missing or inclusive data is marked as "undetermined" in appendix 2.

Although the compendium was created using a systematic process, readers should be aware of its limitations. In the case of multi-site accelerator programs, it is important to note that the inventories contain a mix of accelerator headquarters (or first location) and specific accelerator sites. Some multi-site programs operate continuously at specific sites and some vary the locations of operation. The website provided in the compendium is the primary location website. In addition, the compendium was not systematically reviewed to remove accelerators that are no longer operating. However, in reviewing websites, those accelerators with websites that no longer operated were removed as being defunct.

Approach to Development of Taxonomy

Taxonomies in biology differentiate two types of variation among observed specimens. The first is *genotype*, which refers to fundamentally different characteristics arising from specific variation in their genetic codes. The second is *phenotype*, which refers to observed differences that may arise from the interaction of environmental influences and random genetic variations. A similar approach may be applied to classifying accelerators, incubators, and their variants. Certain characteristics may be viewed as fundamental differences in structure and function, while others may be viewed as minor variances driven by local conditions or personal preferences, for example. Fundamental characteristics that define structure and function are typically evident in an accelerator's business plan and value proposition (*genotype)* variations, while minor variations in service provided among accelerators would be considered *phenotype* variations.

After extensive review of the literature and accelerators listed in appendix 2, Y Combinator stood out as a consistently referenced accelerator, suggesting that there is widespread agreement that Y Combinator exhibits characteristics that tend to define accelerators. Optimal has therefore set Y Combinator's characteristics as an initial benchmark for comparison purposes, recognizing that after the full comparison is complete, the definition of innovation accelerators may vary somewhat from that initial benchmark.[16] Several documents provided background for the development of a generic value proposition and business model to frame this understanding in a structured way (see, for example, Barnes, Blake, and Pinder 2009). In addition, the foundational aspects of accelerators and other related entities were discussed informally with practitioners. Information on observable characteristics was largely obtained through the literature review. After creating the taxonomy flowchart, the classification system was refined using a small random sample of entities from the list in appendix 2.

APPENDIX 2: COMPENDIUM OF ACCELERATOR INVENTORIES

Table A.1. Compendium of Accelerator Inventories

Accelerator	Year founded	City	State	Country	Website	Sources
10Xelerator	2011	Columbus	OH	USA	http://10xelerator.com/	1, 5, 7, 8
500 Startups	2010	Mountain View	CA	USA	http://500.co/	1, 2, 3, 4, 5
Acceleprise	2012	Washington	D.C.	USA	http://www.acceleprise.vc/	1, 3, 5, 7, 8
AlphaLab	2008	Pittsburgh	PA	USA	http://www.alphalab.org	1, 2, 4, 5, 6
Amplify.LA	2011	Los Angeles	CA	USA	http://amplify.la/	1, 4, 5, 7
AngelPad	2010	San Francisco	CA	USA	http://angelpad.org/	1, 3, 4, 5
Berkeley Ventures	2009	Berkeley	CA	USA	http://www.berkeleyventures.com/	4, 5, 7, 8
Betaspring	2009	Providence	RI	USA	http://betaspring.com	1, 2, 4, 5
Bethnal Green Ventures	2010	London	N/A	United Kingdom	http://bethnalgreenventures.com/	1, 3, 7, 8
Bizdom-Detroit	2007	Detroit	MI	USA	http://bizdom.com/	1, 4, 5, 7
Bizdom-Cleveland	2009	Cleveland	OH	USA	http://bizdom.com/	4, 5, 7
Blueprint Health	2011	New York City	NY	USA	http://www.blueprinthealth.org/	1, 2, 3, 4, 5
BoomStartup	2010	Salt Lake City	UT	USA	http://boomstartup.com/	1, 2, 4, 5, 7
Capital Factory	2009	Austin	TX	USA	http://www.capitalfactory.com/	1, 2, 3, 4, 5, 6

Table A.1. (Continued)

Accelerator	Year founded	City	State	Country	Website	Sources
Code for America Accelerator	2009	San Francisco	CA	USA	http://codeforamerica.org/	1, 3, 5, 7, 8
Communitech HYPERDRIVE	2012*	Kitchener	N/A	Canada	http://hyperdrive.communitech.ca/	1, 3, 7
DreamIT Ventures -NYC	2007*	New York City	NY	USA	http://www.dreamitventures.com/	1, 4, 5
DreamIT Ventures - Philadelphia	Undetermined	Philadelphia	PA	USA	http://www.dreamitventures.com/	1, 2, 3, 4, 5, 6
Entrepreneurs Roundtable Accelerator	2011	New York City	NY	USA	http://eranyc.com	1, 2, 3, 4, 5
Environmental Business Cluster	1994	San Jose	CA	USA	http://www.environmentalcluster.org/	3, 4, 7
Extreme Startups	2012	Toronto	N/A	Canada	http://www.extremestartups.com/about/	1, 3, 7, 8
FinTech Innovation Lab	2011	New York City	NY	USA	http://www.fintechinnovationlab.com/	3, 5, 7, 8
Flashpoint	2010	Atlanta	GA	USA	http://flashpoint.gatech.edu/	1, 5, 7, 8
Founder Institute	2009	Silicon Valley	CA	USA	http://fi.co/	3, 4, 7
FounderFuel	2011	Montreal	N/A	Canada	http://founderfuel.com/en/	1, 3, 7, 8
Greenstart	2011	San Francisco	CA	USA	http://greenstart.com/	2, 3, 4, 5

Accelerator	Year founded	City	State	Country	Website	Sources
GrowLab	2011	Vancouver	N/A	Canada	http://www.growlab.ca/	1, 3, 7, 8
Healthbox - Boston	Undetermined	Boston	MA	USA	http://healthbox.com/	1, 3, 5
Healthbox - Chicago	Undetermined	Chicago	IL	USA	http://healthbox.com/	1, 3, 5
Healthbox - London	Undetermined	London	N/A	United Kingdom	http://healthbox.com/	1, 3
H-Farm Ventures	2005	Treviso	N/A	Italy	http://www.h-farmventures.com/en/	1, 2, 7
Hub Ventures	Undetermined	San Francisco	CA	USA	http://www.better.vc/	1, 4, 5
iAccelerator	2008	Ahmedabad	N/A	India	http://iaccelerator.org/	1, 6, 7
ignite100	2011	Newcastle-upon-Tyne	N/A	United Kingdom	http://ignite100.com/	1, 2, 7
Imagine K12	2011	Redwood City	CA	USA	http://www.imaginek12.com/	1, 3, 4, 5, 7
ImpactEngine	2011*	Chicago	IL	USA	http://www.theimpactengine.com/	1, 3, 5, 7
Incubate Miami	2009	Miami	FL	USA	http://www.linkedin.com/company/incubate-miami	1, 5, 7, 8
JOLT	2012	Toronto	N/A	Canada	http://jolt.marsdd.com/	1, 3, 7, 8
Joystick Labs	Undetermined	Durham	NC	USA	http://www.linkedin.com/company/joystick-labs	4, 5
JumpStart Foundry	2010	Nashville	TN	USA	http://www.jsf.co/	1, 4, 5, 7
K5Launch	2011	Los Angeles	CA	USA	http://k5launch.com/	1, 5, 7, 8
Kicklabs(aka Transmedia Capital)	2009	San Francisco	CA	USA	http://transmediacapital.com/	3, 4, 5, 7

Table A.1. (Continued)

Accelerator	Year founded	City	State	Country	Website	Sources
LaunchHouse	2009	Cleveland	OH	USA	http://www.launchhouse.com/	1, 5, 7, 8
Launchpad LA	2009	Los Angeles	CA	USA	http://launchpad.la/	1, 3, 4, 5, 7
Matter.	2012	San Francisco	CA	USA	http://www.linkedin.com/ company/matter-	1, 3, 7, 8
Microsoft Accelerator	Undetermined	Seattle	WA	USA	http://www.microsoft.com/ bizspark/accelerator/	1, 5
Momentum	2008*	Grand Rapids	MI	USA	http://www.momentum.	1, 5, 7
MuckerLab	Undetermined	Santa Monica	CA	USA	http://www.muckercapital. com/	1, 5, 7, 8
NDRC LaunchPad	Undetermined	Dublin	N/A	Ireland	http://www.ndrc.ie/launchpad/	1, 3
NewME Accelerator	2011	San Francisco	CA	USA	http://www.newme accelerator.	4, 5, 7
NYC ACRE	2009	New York City	NY	USA	http://www.nycacre.com/	3, 4, 7
NYC SeedStart	2010*	New York City	NY	USA	http://www.nycseedstart. com/	1, 2, 3, 4, 5, 7
Openfund	2009	Athens	N/A	Greece	http://theopenfund.com/	1, 2, 3, 6, 7
Oxygen Accelerator	2011	Birmingham	N/A	United Kingdom	http://www.oxygen	1, 2, 7
Propeller Venture Accelerator	2010	Dublin	N/A	Ireland	http://www.ryanacademy.ie/ propeller-venture-accelerator	1, 2, 3, 7

Accelerator	Year founded	City	State	Country	Website	Sources
Rock Health – Cambridge	Undetermined	Cambridge	MA	USA	http://rockhealth.com/	3, 5
Rock Health - San Francisco	2011	San Francisco	CA	USA	http://rockhealth.com/	1, 2, 3, 4, 5
Seed Hatchery	2011	Memphis	TN	USA	http://seedhatchery.com/	1, 4, 5, 7
Seedcamp	2007	London	N/A	United Kingdom	http://www.seedcamp.com/	1, 2, 3, 6
SeedRocket	2008*	Barcelona	N/A	Spain	http://www.seedrocket.com/en/	1, 6, 7
StartEngine	Undetermined	Los Angeles	CA	USA	http://www.startengine.com/	1, 5, 8
Startl	Undetermined	New York City	NY	USA	http://startl.org/	1, 2, 3, 7
Startupbootcamp -Copenhagen	2010*	Copenhagen	N/A	Denmark	http://www.startupbootcamp.org/	1, 2, 7
Startupbootcamp -Madrid	2011*	Madrid	N/A	Spain	http://www.startupbootcamp.org/	1, 2, 7
Startup Highway	2011	Vilnius	N/A	Lithuania	http://startuphighway.com/	1, 2, 7
Summer@Highland	2007	Cambridge	MA	USA	http://summer.hcp.com/	4, 5, 7
SURGE Accelerator	2011	Houston	TX	USA	http://surgeaccelerator.com/home	1, 3, 5, 7, 8
Tech Wildcatters	2009	Dallas	TX	USA	http://techwildcatters.com/	1, 2, 3, 4, 5, 7
Techstars - Boston	2009	Boston	MA	USA	http://www.techstars.com/	1, 2, 3, 4, 5, 6
Techstars - Boulder	2007	Boulder	CO	USA	http://www.techstars.com/	1, 2, 3, 4, 5, 6

Table A.1. (Continued)

Accelerator	Year founded	City	State	Country	Website	Sources
Techstars – Cloud	Undetermined	San Antonio	TX	USA	http://www.techstars.com/	1, 3, 5
Techstars - NYC	2011	New York City	NY	USA	http://www.techstars.com/	1, 2, 3, 4, 5
Techstars - Seattle	2010	Seattle	WA	USA	http://www.techstars.com/	1, 2, 3, 4, 5
The Brandery	2010	Cincinnati	OH	USA	http://brandery.org/	1, 2, 3, 4, 5, 7
Think Big Partners	2010*	Kansas City	MO	USA	http://thinkbigpartners.com/	1, 4, 5, 7
University Technology Park	2011	Chicago	IL	USA	http://www.university technologypark.com/	3, 4, 7
Unreasonable Institute	2009	Boulder	CO	USA	http://unreasonableinstitute. org/	3, 4, 7
Victory Spark	Undetermined	Milwaukee	WI	USA	http://gan.co/members/view/vic tory-spark	1, 4
Women Innovate Mobile	2011	New York City	NY	USA	http://wim.co/	1, 4, 5, 7

Accelerator	Year founded	City	State	Country	Website	Sources
Y Combinator	2005	Mountain View	CA	USA	http://ycombinator.com/	1, 2, 3, 4, 5, 6,7

Note: The authors compiled the following compendium (table A.1) from pre-existing lists of accelerators, during early 2014. In the case of multi-site accelerator programs, the entry—including founding year, and location—may reflect the program headquarters, founding location, or one of multiple sites at which the program operates. Optimal did not verify the information outside of the process described in the appendices of this report.[i] The website and source columns were added by Optimal.

[i] Sources: The accelerators in this appendix were compiled from the following six inventories of accelerators and include those accelerators that appear in at least two: (1)Seed-DB, (2) NESTA, (3)Emergent By Design (4) Webbmedia Group, (5)Tech Cocktail, and (6) an early list compiled by Seed-DB founder. Where information about the accelerator was indeterminate (conflicted across sources or was missing), the data were cross-referenced with the following two additional data sources: (7) Crunchbase, or (8) the accelerator's own website or LinkedIn page. Only those accelerators without an active website were removed from the comprehensive list. For more information about the data sources and process of compilation, see the methods section of this report.

* Information was found in only one of the eight sources.

End Notes

[1] The term *innovation ecosystem* refers to clusters and networks of entrepreneurs, inventors, firms, universities, research labs, government agencies, and other institutions, and the resources they bring to bear on new process and product development.

[2] This simplified linear model of the innovation process is presented primarily for the benefit of readers who may be unfamiliar with the process of innovation, and to set up a simple definition of the innovation ecosystem. The actual process of innovation is much more complex, interactive, and variable.

[3] Christiansen subsequently founded the online accelerator database, Seed-DB.

[4] See Seed Database, Seed Accelerator Definition webpage. http://www.seed accessed on August 18, 2014.

[5] Cohen and Hochberg (2014, p. 4). Italics in original

[6] Accelerator subtypes are discussed in more detail in section 4.

[7] Cohen (2013) also discusses this "business model" feature as an important and defining characteristic, however it is not included in the definition proposed in Cohen and Hochberg (2014).

[8] A *value proposition* is a marketing statement that summarizes why a product will solve the potential customer's problem better than another product. A *business model* is the company's plan for how it will generate revenue and make a profit from its operations.

[9] *Rewards* defines the ways in which the products and services described in the activities benefit the customer.

[10] *Value experience* refers to the ways in which potential clients or customers experience and receive the value created by the product—in this case, the accelerator.

[11] *Alternatives* and *differentiation* are intended to identify competitive alternatives to the offered product or service and how the product or service is differentiated from those alternatives.

[12] Cohen (2013) reports the average accelerator program duration to be 3 months and the average equity stake to be 6 percent with a range of 5 percent to 8 percent.

[13] See National Business Incubation Association, What is Business Incubation webpage. https://www.nbia.org/resource_library/what_is/index.php, accessed April 7, 2014.

[14] The Regional Innovation Acceleration Network (RIAN), a project of the State Smart Transportation Initiative funded by the U.S. Economic Development Administration, was founded to establish and support a virtual nationwide community of nonprofit venture development organizations. See Regional Innovation Acceleration Network, Home webpage. http://regionalinnovation.org, accessed August 18, 2014.

[15] See CorpVenturing, Corporate Accelerators webpage. http://www.corpventuring.com/services-corporateaccelerators. html, accessed April 7, 2014.

[16] Because an important reason for defining accelerators is to differentiate them from other models, such as incubators, and to identify variants among accelerators, *relative* differences between operating models determine the import-defining characteristics. Therefore, identification of a preliminary benchmark may be somewhat arbitrary, because it only serves a temporary function.

INDEX

#

20th century, 71
21st century, 2, 18
3D printing, vii, 4, 7, 9, 10, 24, 34, 57, 58, 59, 61, 64, 65, 66, 88

A

access, 4, 17, 18, 19, 26, 27, 28, 29, 33, 34, 35, 37, 40, 41, 42, 43, 48, 51, 54, 70, 71, 76, 78, 82, 83, 86, 94, 99, 102, 105, 112, 124, 128
accessibility, 17, 78
acquisitions, 125
additive manufacturing, vii, 4, 7, 9, 24, 42, 61
administrators, 23, 127
advancement, 7, 31
advocacy, 3, 57
aerospace, 8, 11, 30, 86
Africa, 8
agencies, 3, 18, 24, 27, 40, 44, 47, 48, 52, 54, 55, 56, 64, 65, 67, 83, 142
agility, 75
alternative energy, 3
arbitrage, 78
ARC, 61
Asia, 8, 101
Asian countries, 23
assessment, 40, 95, 104
assets, 34, 36, 42, 64, 70, 71, 88, 113, 127
attribution, 50
authority, 48, 52, 56
automakers, 72, 80, 81
automation, 12, 13, 30
automobile(s), 72, 81
aversion, 47
awareness, 41

B

background information, 20
bank financing, 34
bankruptcy, 17
banks, 16, 35, 36
bargaining, 79
barriers, vii, 2, 3, 4, 7, 11, 13, 14, 18, 19, 20, 40, 41, 43, 55, 62, 70, 74, 77, 81, 102
base, 16, 70, 83, 84, 96
basic research, 17, 22, 23, 24, 26, 28, 62
benchmarking, 91
benefits, 6, 22, 31, 32, 40, 41, 42, 50, 65, 79, 81, 85, 89, 106, 109, 121, 124
bias, 104
biotechnology, 37
blogger, 132
blogs, 132
boot camps, vii, 94, 99
borrowers, 20, 21, 22

Brazil, 8, 23, 61
broadband, 17
Bureau of Labor Statistics, 87
business environment, 54
business model, 8, 9, 30, 42, 44, 75, 94, 96, 101, 106, 107, 110, 111, 112, 114, 115, 116, 117, 118, 119, 120, 126, 127, 128, 134, 142
business startup, vii, 94
businesses, vii, 1, 2, 3, 4, 5, 6, 7, 10, 14, 15, 17, 18, 28, 30, 32, 33, 34, 35, 36, 38, 41, 44, 45, 46, 48, 52, 54, 55, 57, 62, 71, 83, 95, 98, 99, 101, 102, 105, 108, 121, 125
buyer(s), 80, 82, 84, 90

C

Cabinet, 66, 67
CAD, 8, 11, 29, 30, 50
cadmium, 79
campaigns, 34
candidates, 38
capital expenditure, 86
capital intensive, 34
capital markets, 16
carbon, 74, 80
carbon monoxide, 80
cash, 26, 35, 108, 122
cash flow, 26, 35
casting, 71
catalyst, 80, 81
Census, 5, 72, 73, 74, 75, 87, 88, 89, 125
ceramic, 12
ceramic materials, 12
certification, 37
Chad, 89
challenges, 2, 3, 4, 7, 11, 12, 17, 18, 19, 20, 31, 32, 33, 34, 39, 47, 55, 56, 67, 70, 71, 76, 77, 82, 86, 128
Chicago, 25, 137, 140
Chile, 101
China, 8, 23, 53, 63, 78
cities, 101, 102
citizens, 47
clarity, 94, 97
classification, 97, 100, 106, 107, 120, 134
clean energy, 25, 99
clients, 112, 142
climate, 17
clusters, 40, 83, 142
CNN, 60
coal, 72
collaboration, 23, 24, 40, 78, 81, 85
collateral, 34, 35, 36
colleges, 22, 30, 31, 32, 33, 63
commerce, 89
commercial, 22, 27, 39, 66, 108, 109
communication, 79, 81
community(s), 30, 31, 33, 35, 54, 63, 83, 86, 142
compatibility, 47
compensation, 20, 33
competition, 29, 31
competitive advantage, 114
competitiveness, vii, 56, 67, 70, 73, 74, 77, 84
competitors, 16, 23, 70, 77, 80
compilation, 141
complement, 10, 85
complexity, 11, 13, 30, 46, 47, 56
compliance, 30, 53
composites, 12, 25
computer, 8, 12, 13, 63, 72
computer software, 13
computing, 34
conflict, 38
Congress, 57
consensus, 29, 94, 133
consolidation, 10
consulting, 85, 99
consumers, 9
containers, 84
convergence, 103
conversations, 27
cooling, 12
coordination, 44, 56, 78, 80, 81, 109
cost, 8, 9, 10, 11, 13, 14, 15, 41, 44, 45, 46, 52, 76, 77, 78, 79, 80, 82, 86, 89, 90, 91, 102, 104, 127
cost-benefit analysis, 41

counsel, 3, 60
counterfeiting, 50, 65
creativity, 75
criticism, 35, 94, 104, 105
currency, 80
curricula, 32
curriculum, 27, 63
customer relations, 55
customer service, 54
customers, 11, 13, 26, 41, 42, 54, 79, 83, 90, 99, 107, 111, 114, 122, 126, 142
cycles, 78

D

data collection, 124, 127, 128
database, 40, 44, 48, 103, 124, 125, 132, 133, 142
debentures, 38, 64
defects, 89
Denmark, 139
Department of Agriculture, 25
Department of Commerce, 52, 54, 66, 70, 89
Department of Defense, 40, 48, 83, 90
Department of Energy, 25, 40, 48, 71, 83
Department of Homeland Security, 48
Department of Labor, 31, 63
Department of Transportation, 48
deposition, 9, 65
depth, 55
designers, 9
developing countries, 2
diffusion, vii, 19, 41, 42, 43, 44, 69, 70, 74, 75, 76
digital manufacturing, 12, 13, 14, 30, 33, 45, 57
directors, 105
distribution, 6, 103, 114, 122, 128
division of labor, 88

E

economic development, 6, 15, 24, 54, 83, 100, 112, 113, 115, 118
economic growth, 96
economics, 121
economies of scale, 79
ecosystem, 2, 3, 4, 16, 17, 18, 28, 56, 57, 59, 64, 79, 85, 95, 100, 106, 109, 114, 125, 130, 132, 142
education, 17, 20, 24, 29, 31, 56, 61, 63
educational institutions, 24
educational programs, 59
Egypt, 8
electricity, 71
e-mail, 124
empirical studies, 96
employees, 5, 6, 15, 16, 29, 30, 60, 66, 71, 73, 75, 86, 87, 122, 124
employers, 31, 32, 63
employment, vii, 7, 31, 62, 73, 83, 125
energy, 44, 81, 84, 87
energy consumption, 44
energy efficiency, 81
enforcement, 65
engineering, 1, 16, 18, 29, 31, 32, 43, 70, 74, 76, 81, 90
enrollment, 21
entrepreneurs, vii, 1, 2, 4, 16, 26, 27, 28, 37, 39, 40, 49, 55, 59, 62, 94, 95, 98, 99, 105, 106, 108, 112, 113, 114, 142
entrepreneurship, 20, 21, 96, 125, 132
environment(s), 6, 10, 49, 56, 78, 85
environmental influences, 134
equipment, 12, 13, 19, 22, 33, 39, 40, 43, 45, 46, 47, 53, 70, 71, 76, 77, 79, 82, 86, 112
equity, 16, 34, 35, 37, 38, 64, 94, 98, 99, 100, 101, 103, 108, 110, 111, 113, 115, 116, 118, 120, 122, 126, 142
Europe, 8, 91, 102
European Union, 102
evidence, 9, 65, 88, 89, 90, 128
exclusion, 127
executive branch, 67

expenditures, 63
expertise, 2, 28, 39, 43, 51, 52, 64, 66, 70, 76, 77, 79, 81, 82, 84, 111
export control, 50, 53, 66
export market, 53
export promotion, 53
exporters, 52
exports, 52, 54, 66, 67
external environment, 13
external financing, 99
external relations, 5

F

Fabrication, 57
factories, 13, 72
families, 20
fear, 41, 56, 70
federal agency, 67
federal government, 1, 2, 4, 5, 16, 17, 18, 22, 23, 44, 47, 53, 62, 67, 84, 128
Federal Reserve, 62
Federal Reserve Board, 62
fiber, 74
financial, 17, 32, 34, 35, 36, 40, 41, 52, 74, 86, 112, 115, 118, 122, 124, 125
financial capital, 115
financial performance, 34, 35
financial resources, 32, 40
financial support, 17, 52, 86
firearms, 66
firm management, 40, 41
firm size, 89
fiscal year, 22
flexibility, 48, 76, 113, 120
fluctuations, 80
fluid, 18
force, 66
Ford, 71, 87
Ford, Henry, 87
foreign investment, 54
formation, 35
foundations, 109
France, 8, 53
freedom, 10
funding, 17, 18, 22, 23, 27, 28, 31, 33, 34, 36, 37, 38, 41, 49, 51, 62, 64, 83, 85, 94, 95, 98, 100, 101, 108, 109, 113, 115, 118, 120, 122, 123, 124, 125, 127, 133
fundraising, 37
funds, 34, 38, 56, 64, 86, 99, 102

G

GAO, 65
GDP, 62
General Motors, 89
genetic code, 134
genotype, 134
Germany, 8, 53, 85, 86, 91
global economy, 2, 18
global markets, 43
goods and services, 46
government spending, 17
governments, 100, 128
GPS, 17
grants, 17, 54, 64, 109, 113, 125
Great Recession, 71
Greece, 138
growth rate, 8, 23, 61
guidance, 27, 78, 112
guidelines, 43, 124

H

health, 73, 118
health information, 118
hiring, 15, 111
history, 34, 55
hobby, 45
host, 102
House, 25, 51, 62, 67
House of Representatives, 25, 51, 62
human, 108
hydrocarbons, 80

I

IDA, 59, 62

identification, 61, 100, 133, 142
images, 57
immigration, 20
improvements, 8, 20, 65, 76, 80, 82, 90
income, 21, 26, 62, 64, 110
income tax, 62
incubator, 98, 101, 103, 104, 112, 115, 119, 120, 121, 132, 133
India, 8, 23, 137
individuals, 2, 3, 4, 9, 13, 16, 20, 27, 28, 29, 30, 31, 33, 37, 45, 50, 57, 123
Indonesia, 24
inflation, 29, 87
information technology, 31
infrastructure, 13, 14, 17, 20, 24, 42, 56
injury, 50
innovative products, vii, 1, 23, 46, 48, 76
institutions, 16, 24, 56, 57, 86, 99, 142
integration, 13
intellectual property, 17, 26, 34, 58, 65, 80, 99, 115
intellectual property rights, 17, 65
intelligence, 82
intermediaries, 83
internal rate of return, 122
international competitiveness, 5
intervention, 123
inventors, 52, 114, 142
investment(s), 7, 15, 27, 34, 35, 36, 37, 38, 39, 41, 51, 60, 64, 69, 71, 74, 76, 77, 78, 79, 81, 84, 89, 94, 96, 99, 102, 103, 105, 106, 108, 109, 112, 113, 115, 118, 119, 112, 124, 126
investment bank, 124
investors, 4, 16, 34, 36, 37, 38, 51, 64, 94, 102, 108, 109, 122, 123, 124, 125, 126
IPO, 93, 108
IPR, 17
Ireland, 63, 138
iron, 71
IRS, 25
Israel, 8
issues, 4, 12, 13, 14, 16, 19, 20, 29, 37, 45, 47, 49, 50
Italy, 137

J

Japan, 8, 63
job creation, 5, 15, 87, 104, 111, 118, 126
job skills, 17
job training, 31, 63
job training programs, 31, 63

L

labor market, 30
landscape, 44, 127
laws, 17, 18, 35, 44, 49, 53
laws and regulations, 18, 44, 49, 53
lawyers, 18, 36
lead, 2, 7, 10, 11, 27, 29, 33, 36, 39, 40, 44, 51, 72, 73, 76, 77, 78, 79, 80, 81, 85, 88, 89, 90, 95, 120
leadership, 17, 53
learning, 26, 27, 41, 81
legal issues, 49
legislation, 32, 37, 50, 51
lending, 36
light, 23, 71, 81, 126
limestone, 72
linear model, 142
Lithuania, 139
litigation, 51
loans, 20, 21, 22, 35, 36, 62
local community, 99
local conditions, 125, 134
local government, 31, 83
logistics, 8
love, 121
low temperatures, 84

M

major issues, 20
majority, vii, 1, 10, 32, 69, 73, 96
management, 14, 34, 38, 39, 55, 60, 64, 82, 99, 112
manufacturing, vii, 3, 4, 6, 7, 8, 9, 10, 11, 12, 13, 14, 15, 24, 25, 29, 30, 31, 32, 33,

40, 42, 43, 45, 50, 57, 61, 63, 65, 66, 69, 70, 71, 72, 73, 74, 75, 76, 77, 78, 79, 82, 83, 84, 87, 88, 96, 128
manufacturing companies, 13, 75
mapping, 83
market failure, 96, 128
market share, 16, 89
marketing, 5, 15, 54, 60, 76, 80, 100, 142
marketplace, 17, 95, 98, 100, 107, 108, 113, 121
Maryland, 34, 35, 60, 64
MAS, 85
materials, 4, 7, 8, 10, 11, 12, 13, 14, 25, 33, 42, 43, 46, 53, 66, 74, 77, 81, 87
matter, 75, 138
measurement, 104
media, 7, 57, 90, 97, 132
medical, 8, 14, 99
medicine, 17
membership, 3, 98
mentor, 27
mentoring, vii, 78, 94, 98, 103, 106, 108, 109, 111, 113, 114, 123
mentorship, 85, 103, 105, 120, 126
mergers, 125
meta-analysis, 89
metals, 11, 12, 42, 61, 74, 79, 80, 81
methodology, 96
Mexico, 72, 83, 84, 90
Miami, 137
Microsoft, 61, 105, 114, 138
military, 21, 66, 83
mission(s), 4, 24, 47, 60, 83
mobile device, 96
models, 8, 31, 32, 44, 63, 119, 127, 128, 142
momentum, 138
motivation, 94, 112, 115, 120, 126
multinational corporations, 2

N

nanotechnology, 99
National Economic Council, 58
National Institutes of Health, 27, 48, 67
National Research Council, 59, 60, 91
natural resources, 72
networking, 23, 85, 101, 108, 113, 114
New England, 113
New York Stock Exchange, 125
niche market, 109
nodes, 63
nonprofit organizations, 24, 100, 107, 128
null, 109

O

Obama, President Barack, 2, 7, 21, 24, 25, 31, 51, 52, 56, 66, 90
obstacles, 65, 75
OECD, 6, 58, 60, 65
Office of Advocacy, vii, 1, 2, 3, 5, 55, 57, 59, 60, 62, 93
Office of Management and Budget, 56, 65
officials, 4
Oklahoma, 113
open markets, 54
operations, 30, 44, 49, 61, 79, 83, 89, 101, 142
opportunities, 14, 17, 18, 19, 20, 22, 23, 27, 40, 44, 46, 47, 48, 57, 58, 78, 82, 84, 86, 99, 105, 108, 113, 115, 125, 128
opportunity costs, 108
optimization, 11
organ, 115
organic growth, 5
organize, 107
outreach, 2, 3, 4, 7, 32, 54, 55, 57
outsourcing, 71
overlap, 122
Overseas Private Investment Corporation, 52
oversight, 56, 80
ownership, 82
oxidation, 80
oxygen, 138

P

parity, 85
participants, 2, 17, 18, 23, 31, 33, 49, 50, 118, 122, 123, 128
patents, 26, 28, 49, 51, 65, 66
payroll, 25, 26, 62, 125
permission, 48
pharmaceuticals, 16, 37
phenotype, 134
Philadelphia, 101, 136
pipeline, 94, 102
pitch, 103, 120, 126
plants, 79
plastics, 11, 42
platform, 11, 28, 33
playing, 66
policy, 4, 20, 96, 97, 99, 100, 104, 123, 126, 127, 132
policymakers, vii, 4, 50, 105, 121, 127
pollination, 70, 77
poor performance, 34
population, 104
portfolio, 106, 112, 115, 121, 122
potential benefits, 89
president, 59, 90
President, v, 2, 7, 13, 21, 23, 24, 25, 31, 35, 51, 52, 54, 56, 60, 66, 69, 88, 90
private good, 115, 118
private sector, 6, 17, 22, 32, 43, 46, 47, 54, 59, 62, 64, 71, 74, 77, 84, 100, 127
probability, 115, 121
problem solving, 30
problem-solving, 79, 89
problem-solving behavior, 89
procurement, 46, 47, 48, 65
producers, 9
product design, 10, 14, 15, 39, 80
production costs, 6, 12
professionals, 31, 41
profit, 16, 24, 31, 59, 91, 95, 100, 104, 106, 107, 110, 111, 113, 115, 118, 119, 120, 121, 126, 127, 142
project, 27, 37, 64, 83, 85, 123, 128, 133, 142
proposition, 94, 107, 109, 111, 113, 115, 117, 120, 126, 134, 142
prosperity, 1
protection, 15, 36, 50
prototype(s), 9, 14, 48, 64, 76, 128
public policy, 106, 127
public resources, 95
public sector, 70, 81, 95
public service, 106
publishing, 44
Puerto Rico, 43, 63

Q

quality of life, 1
quality standards, 13

R

race, 70, 77
radio, 61
rate of return, 121, 122
raw materials, 72
real time, 13
reality, 81
recidivism, 103
recidivism rate, 103
recommendations, 4, 56, 63
recovery, 73
reform, 36, 56, 62, 66, 67
regional economies, 112
regulations, 48, 49, 127
relevance, vii, 132
relief, 21
renaissance, 84
repair, 14
reputation, 108
requirement, 36, 38
requirements, 11, 30, 32, 35, 45, 48, 50, 53, 54, 107, 110, 128
researchers, 4, 16, 18, 23, 26, 27, 39, 61, 85, 121, 125, 127, 128
resolution, 94
resource management, 108

resources, vii, 2, 14, 16, 22, 24, 27, 28, 29, 31, 39, 40, 52, 55, 56, 57, 72, 74, 77, 85, 86, 91, 94, 112, 116, 117, 122, 128, 129, 131, 142
response, 128
responsiveness, 89
restrictions, 53, 66
retail, 14
revenue, 5, 8, 15, 22, 49, 63, 94, 110, 122, 142
rewards, 108, 109
rights, 17, 49, 50, 115
risk(s), 16, 22, 27, 34, 36, 39, 46, 47, 53, 56, 70, 76, 78, 79, 80, 83, 109, 111
risk taking, 56
roots, 96
royalty, 38
rubber, 72
rubber products, 73
rules, 10, 35, 43, 115
Russia, 8

S

safety, 43, 49
Saudi Arabia, 129
savings, 64
scaling, 19
school, 21, 33, 86
science, 1, 29, 32, 60, 99
scope, 31, 50, 76
security(s), 13, 14, 35, 36, 38, 39, 45, 61, 66, 78
seed, 34, 36, 60, 94, 95, 99, 103, 105, 106, 107, 108, 113, 114, 115, 116, 120, 126, 130, 142
selectivity, 127
semiconductors, 72
Senate, 25, 32, 37, 62
sensors, 13, 74, 79
September 11, 57
service provider, 9, 61, 124
shape, 11, 12, 57, 88, 126
shortage, 18, 29, 30, 43
short-termism, 90
showing, 7
Silicon Valley, 29, 136
silver, 79
Singapore, 63
skilled workers, 29, 32
skills training, 123
small business(s), 3, 4, 5, 7, 9, 13, 14, 15, 16, 17, 18, 19, 22, 25, 29, 30, 32, 33, 35, 38, 40, 41, 43, 44, 45, 46, 47, 48, 49, 50, 52, 53, 54, 55, 64, 65, 70, 75, 78, 81, 82, 83, 84, 90, 102, 123
small firms, vii, 14, 25, 33, 42, 45, 46, 48, 50, 52, 54, 65, 70, 75, 77, 86, 88, 89
small manufacturers, vii, 30, 69, 70, 71, 73, 74, 75, 76, 77, 81, 82, 84
social capital, 109, 115, 127
social network, 99, 127
society, 1, 27, 50
software, 8, 10, 12, 13, 29, 30, 33, 45, 55, 94, 96, 99, 102, 110
solution, 15, 28, 39, 79, 86
South Africa, 8
South America, 101
soybeans, 71
Spain, 139
specialization, 111
species, 107
spending, 22, 23, 62, 83
spillovers, 88
spreadsheets, 124
stability, 66
staffing, 117
stakeholders, 3, 51
startup assistance programs, vii, 94, 106, 127
startup firm, 122
state(s), 6, 13, 16, 21, 23, 31, 35, 40, 43, 51, 54, 63, 64, 70, 71, 82, 83, 91, 100, 102, 133
State of the Union address, 2
statistics, 61, 87, 124
steel, 71, 81
stimulation, 42
stock, 35, 125
storage, 13, 14, 45, 80, 84

stress, 69
structural characteristics, 126
structure, 40, 52, 102, 115, 134
supplier(s), vii, 42, 69, 70, 71, 72, 73, 74, 75, 76, 77, 78, 79, 80, 81, 82, 84, 85, 86, 88, 89, 90
supply chain, vii, 7, 8, 13, 14, 66, 69, 70, 71, 73, 74, 76, 77, 78, 80, 81, 82, 83, 84, 85, 89
support services, 31, 95, 112
survival, 104, 123

T

talent, 16, 18, 19, 23, 29, 30, 43
target, 100, 103
Task Force, 87, 88
tax credits, 37, 63
tax incentive, 26, 33, 63
tax policy, 17
tax rates, 26, 63
taxes, 25, 26
taxonomy, 94, 95, 96, 106, 107, 120, 134
team members, 27
teams, 27, 40, 99, 103, 114
technical assistance, 79, 80, 85, 86, 106, 108, 111, 112, 113, 114, 126
technical change, 75
technician, 30
technological advances, 77
technological developments, 77
technology industries, vii, 4, 7, 52, 109
technology transfer, 43, 63, 113
temperature, 42
testing, 14, 15, 39, 100
textiles, 72
threats, 65
time frame, 106, 121
total product, 10
Toyota, 89
toys, 65
trade, 4, 18, 54, 67, 85
training, 17, 21, 22, 24, 27, 29, 30, 31, 33, 40, 51, 54, 55, 56, 76, 77, 84, 85, 90, 106
training programs, 30, 31
translation, 54
transmission, 71
transparency, 66
transportation, 17, 52, 80
Transportation Security Administration, 48

U

U.S. Department of Commerce, v, 58, 69, 89
U.S. economy, 44, 84, 96
U.S. manufacturing, vii, 6, 24, 69, 71, 72, 73, 74, 84
U.S. Small Business Administration, vii, 1, 60, 62, 93
underwriting, 35, 36
unions, 31, 85
United Kingdom, 85, 91, 101, 130, 135, 137, 138, 139
United States, vii, 1, 2, 5, 6, 7, 8, 9, 16, 17, 18, 22, 23, 24, 25, 30, 31, 56, 58, 60, 62, 64, 71, 74, 79, 83, 95, 96, 101, 102, 113, 129
universities, 4, 16, 18, 22, 24, 31, 32, 63, 83, 85, 86, 100, 106, 113, 142
USA, 13, 135, 136, 137, 138, 139, 140, 141

V

validation, 99
valuation, 36
variables, 124
variations, 9, 10, 111, 132, 134
venture capital, 4, 16, 18, 34, 60, 94, 102, 124, 132
vertical integration, 71
Vice President, 31, 63
Volkswagen, 114
vouchers, 83, 85, 86

W

Washington, 59, 60, 61, 135
waste, 90

weakness, 122
web, 87, 88
websites, 44, 47, 54, 63, 132, 133, 134
welding, 63, 79
well-being, 120
White House, 6, 10, 57, 60, 61, 62, 63, 66, 67, 87, 90, 95
Wisconsin, 78, 90
work environment, 40
workers, 13, 29, 30, 31, 33, 72, 73, 83, 85
workforce, 13, 17, 24, 29, 32, 43, 56, 63, 73, 75
working groups, 51
workplace, 5
World Bank, 58
worldwide, 8

Y

yield, 64, 106